STATE OF THE POLAR BEAR REPORT 2017

Susan J. Crockford

ISBN 978-0-9931190-1-9
Printed by Createspace.
Cover image ©Shutterstock.

Published by:
The Global Warming Policy Foundation
55 Tufton Street
London SW1P 3QL

www.thegwpf.org

Contents

About the author		**iv**
Foreword		**iv**
Executive summary		**v**
1	**Introduction**	**1**
2	**Conservation status**	**2**
3	**Population size**	**3**
	Global	3
	Subpopulations by ecoregion	3
4	**Population trends**	**12**
	The problem of statistical confidence	14
5	**Habitat status**	**14**
	Global sea ice	14
	Sea ice loss by subpopulation	18
	Freeze-up and breakup date changes	18
6	**Prey base**	**19**
7	**Health and survival**	**22**
	Body condition	22
	Effect of record low winter ice	23
	Hybridization	23
	Effect of contaminants	24
	Cannibalism	24
	Den collapses	24
	Swimming bears	24
	Denning on land	25
	Ice-free period on land	25
	Litter sizes	25
8	**Evidence of flexibility**	**25**
	Den locations	25
	Feeding locations	27

Genetics 28

9 **Human–bear interactions** **29**
Attacks on humans 29
Problem bears in winter/spring 29

10 **Discussion** **31**

Notes **34**

Bibliography **35**

About the GWPF **54**

About the author

Dr Susan Crockford is an evolutionary biologist and has been working for 35 years in archaeozoology, paleozoology and forensic zoology.[1] She is an adjunct professor at the University of Victoria, British Columbia, but works full time for a private consulting company she co-owns (Pacific Identifications Inc). She is the author of *Rhythms of Life: Thyroid Hormone and the Origin of Species*, *Eaten: A Novel* (a polar bear attack thriller), *Polar Bear Facts and Myths* (for ages seven and up, also available in French and German), *Polar Bears Have Big Feet* (for preschoolers), and the fully referenced *Polar Bears: Outstanding Survivors of Climate Change,*[2] as well as a scientific paper on polar bear conservation status.[3] She has authored several earlier briefing papers and videos for GWPF on the subjects of polar bears and walrus.[4] Susan Crockford blogs at www.polarbearscience.com.

Foreword

From 1972 until 2010,[5] The Polar Bear Specialist Group (PBSG) of the International Union for the Conservation of Nature (IUCN) published comprehensive status reports every four years or so, as proceedings of their official meetings, making them available in electronic format. Now, the PBSG disseminates information only on its website, updated (without announcement) at its discretion. But none of the information is compiled into a standalone document for archival purposes and new data replaces old without a copy of the replaced information sent to an archive.

This State of the Polar Bear Report is intended to provide the kind of content once available in those PBSG meeting reports, albeit with more critical commentary regarding some of the inconsistencies and sources of bias present in the corpus of reports and papers. It is a summary of the current state of polar bears in the Arctic since 2014, relative to historical records, based on a review of the recent and historical scientific literature. It is intended for a wide audience, including scientists, teachers, students, decision-makers and the general public interested in polar bears and Arctic ecology.

Executive summary

- Global polar bear numbers have been stable or risen slightly since 2005, despite the fact that summer sea ice since 2007 hit levels not expected until mid-century: the predicted 67% decline in polar bear numbers did not occur.
- Abundant prey and adequate sea ice in spring and early summer since 2007 appear to explain why global polar bear numbers have not declined, as might have been expected as a result of low summer sea ice levels.
- The greatest change in sea ice habitat since 1979 was experienced by Barents Sea polar bears and the least by those in Southern Hudson Bay, the most southerly region inhabited by bears.
- As far as is known, the record low extent of sea ice in March 2017 had no impact on polar bear health or survival.
- Some studies show bears are lighter in weight than they were in the 1980s, but none showed an increase in the number of individuals starving to death or too thin to reproduce.
- A just-released report of Southern Beaufort Sea bears having difficulty finding prey in 2014–2016 suggests that the thick ice events that have impacted the region every ten years or so since the 1960s have continued despite reduced summer sea ice.
- Claims of widespread hybridization of polar bears with grizzlies were disproven by DNA studies.
- Overly pessimistic media responses to recent polar bear issues have made heartbreaking news out of scientifically insignificant events, suggesting an attempt is being made to restore the status of this failed global warming icon.

1 Introduction

The US Geological Survey estimated the global population of polar bears at 24,500 in 2005.[6] In 2015, the International Union for the Conservation of Nature (IUCN) Polar Bear Specialist Group (PBSG) estimated the population at 26,000 (range 22,000–31,000)[7] but additional surveys published since then bring the total to near 28,500 with a relatively wide margin of error.[8] This is the highest it has been since the bears were protected by international treaty in 1973.[9] While potential measurement error means it can only be said that the global population has likely been stable since 2005 (but may have increased slightly), it is far from the precipitous decline experts expected given summer sea ice levels as low as they have been in recent years.[10]

Between 2007 and 2015, summer sea ice on average dropped about 38% from 1979 levels, an abrupt decline within measurement error of the reduced coverage expected to occur by mid-century (Figure 1).[11] Christine Hunter and colleagues[12] proclaimed in 2007 that such reduced summer sea ice by 2050, if present for eight out of ten years (or four out of five years), would generate a massive drop in polar bear numbers: 10 vulnerable subpopulations out of 19 would be extirpated, leaving fewer than 10,000 animals worldwide (a 67% decline). Fortunately, recent research shows this decline in polar bear abundance has not occurred, which indicates summer sea ice levels are not as critical to polar bear survival as USGS biologists assumed.[13]

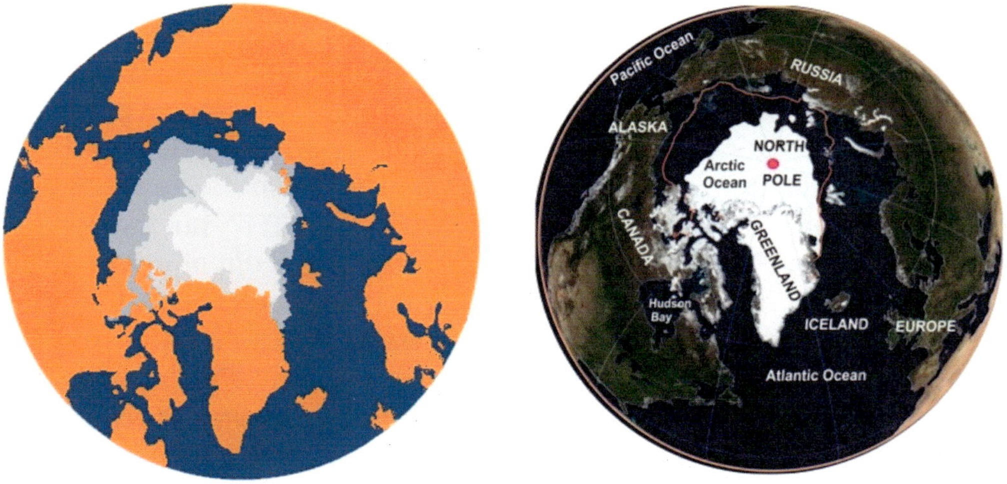

(a) (b)

Figure 1: Predicted sea ice changes versus observations.
(a) Predictions, based on 2004 data, for 2020 (light grey), 2050 (mid grey), and 2080 (dark grey)(ACIA 2005). These were used in 2007 to predict a 67% decline in global polar bear numbers. (b) Situation as at 10 September 2012, an example of sea ice extent experienced since 2007. See Crockford 2017a for details.

Some recent studies show declines in average weights of polar bears compared to the 1980s, but none recorded an increase in the number of individuals starving to death or too thin to reproduce.[14] Although some photos of starving bears have garnered media attention, most bears have been found to be in good-to-excellent condition. In fact, photos of fat bears seem to outnumber those of thin bears in recent years.

Studies on harp seals in the Davis Strait, and on ringed and bearded seals in the Chukchi Sea, show these Arctic seals are doing much better than they were in the 1980s, despite large local declines in sea ice, indicating the critical prey base for polar bears in these regions has improved with less ice. The same is likely true in other regions that have not been studied (except for the Southern Beaufort), although this is often left unsaid. Overall, abundant prey and sea ice in spring and early summer since 2007 appears to explain why polar bear numbers have not declined as might have been expected as a result of low summer sea ice levels.

Overly pessimistic media responses to recent polar bear issues, however, have been out of proportion to their scientific significance. It is not just the starving polar bear images splashed across the internet, but the lacklustre research results promoted beyond reason.[15] It is almost as if the media and the people who feed them are actively trying to restore the polar bear to its former glory as beloved icon of anthropogenic global warming, a tactic that seems destined to backfire.

2 Conservation status

The IUCN, in their 2015 Red List assessment, again listed the polar bear as 'vulnerable' to extinction, as it did in 2006.[16] Similarly, in 2016, the US Fish and Wildlife Service upheld its 2008 conclusion that polar bears were 'threatened' with extinction under the US Endangered Species Act (ESA). In both of these instances, polar bear conservation status is based on computer-modeled future declines predicted to exceed standard threshold levels (i.e. a population decline of 30% or more expected within three generations), not observed declines. The species was listed as one of 'Special Concern' by Environment Canada in 2008 and 2011[17] but the population status maps published online in 2014 are now out of date.[18] Polar bears currently have a relatively large population size and their historical range has not diminished due to habitat loss since 1979. If assessed on current observations, the polar bear would qualify for a status of 'Least Concern' in the IUCN Red List in 2015 (as the species would have done in 2006 too) and the ESA would not have included polar bears on its list of threatened and endangered species in 2008.[19] Thus, concerns about the conservation status of polar bears are all about 'the potential response of the global population of polar bears to projected sea ice declines',[20] not their current population size.

The IUCN (2006) and ESA (2008) assessments of the polar bear were the first time these organisations had used predicted population declines based on climate models; all other species (with only a few recent exceptions) are assessed based on population declines already observed. As a consequence, the public and the media often logically assume that polar bear numbers must be currently declining because they have been listed as 'threatened' or 'vulnerable' – because this would be true for all other species listed by the IUCN or the ESA, with only a few exceptions. This confusion is understandable because it appears contradictory. But the peculiar way in which polar bear conservation status has been defined means it is entirely correct to state that polar bears are currently thriving, and to insist that such a statement is not at odds with a conservation status based on possible *future* declines in population size.

3 Population size

Global

Despite the fact that one of the primary objectives of the government-appointed PBSG members in 1973 was to generate a global population estimate,[21] this portion of their mandate has proven particularly difficult to attain. Despite more than 50 years of dedicated research, several subpopulations have never been comprehensively surveyed for population abundance (East Greenland, Chukchi Sea, Arctic Basin, Laptev Sea) and several others have had only one survey conducted over that period (Kara Sea, Viscount Melville, Lancaster Sound, McClintock Channel, and Norwegian Bay, although as of late 2017, surveys for Viscount Melville and McClintock Channel were in progress).

In 1993, the PBSG estimated polar bear abundance at about 21,470–28,370 (a figure rounded to 22,000–27,000 in 1997). This number was 'adjusted' to 21,000–25,000 in 2001 and 'further simplified' to 20,000–25,000 in 2005 (where the apparent decline since 1993 comes from the fact that some estimates used prior to 2001 were deemed to be not scientific enough and were dropped from the totals).[22] In contrast, in 2005 the US Geological Survey put the global population of polar bears at 24,500, a mid-point estimate used to support the US Fish and Wildlife Endangered Species Act listing in 2008.

In 2014, the PBSG mid-point estimate was listed as 'approximately 25,000' (no range given), which was still the figure listed on their website at 17 January 2018. This is rather odd, since the 2015 IUCN Red List assessment, written by PBSG members,[23] used a mid-point estimate of 26,000 (but not 26,500, the true mid-point of the stated 22,000–31,000 range, apparently due to potential estimate errors).

However, additional survey results published since the 2015 Red List assessment was prepared should bring the mid-point total at 2015 to near 28,500, with a similarly wide margin of error (see Section 4 for more detail). Note that a new estimate for Gulf of Boothia, Viscount Melville, McClintock Channel, and Southern Hudson Bay are pending as of 2017, which could put that global mid-point estimate even higher. The abrupt drop in summer sea ice that occurred in 2007[24] was not predicted by experts to occur until mid-century[25] yet the predicted decimation of polar bears worldwide expected under those conditions (a loss of two thirds of the global total, leaving only 6,660–8,325 bears remaining) did not come even close to happening.[26]

Subpopulations by ecoregion

The US Geological Survey defined four Arctic sea-ice ecoregions in 2007 as part of their current and future assessments of polar bear population size and health (Figure 2).[27]

- The 'Seasonal' ecoregion represents all the subpopulation regions where sea ice melts completely during the summer, stranding polar bears onshore.
- The 'Divergent' ecoregion includes all subpopulation regions where sea ice recedes from the coast into the Arctic Basin during the summer, leaving bears the option of staying onshore or remaining with the sea ice

- The 'Convergent' ecoregion is the subpopulation regions where ice formed elsewhere drifts towards shore all year long.
- The 'Archipelago' ecoregion is the subpopulations in the Canadian Arctic archipelago.

The ecoregion concept now appears to have been accepted as a useful assessment methodology.[28] However, a strong argument could be made that the Southern Beaufort (SB; in the 'Divergent' ecoregion) should be treated as an outlier to this classification. Its unusual status is highlighted by the fact that it is the only subpopulation among 19 worldwide that was considered 'declining' in 2014 (see Section 4). All evidence for this declining status points to thick spring ice conditions as the cause. Such conditions in SB are episodic, occurring every ten years or so and persisting for 2–3 years. The most devastating and well-documented thick spring-ice events occurred in 1974–1976 and 2004–2006,[29] with evidence of perhaps less severe events in the early 1960s, mid-1980s, early 1990s, and the mid-2010s.[30] Such conditions (as far as is known) are unique to this one Arctic region and periodically have a severe impact on polar bear health and survival, and thus population size: nothing remotely like it affects the (adjacent) Chukchi Sea or Barents Sea subpopulations, which are also classified as 'divergent' sea-ice ecotypes. Canada has proposed a change in the boundary between the Southern and Northern Beaufort regions (Figure 3). This will make management easier, but if the changes are adopted by the PBSG, accurately tracking long-term changes in population size and effects of thick spring ice events will become extremely difficult. Most of the population declines that occurred every ten years or so in the past were movements of bears away from thick ice areas, rather than actual deaths and the boundary change means these movements will no longer happen within the SB but between SB and NB.

Baffin Bay – Seasonal

A comprehensive survey of Baffin Bay (BB) polar bears undertaken in 1993–1997[33] generated an estimate of 2,074±226.[34] The government report for the latest survey, completed in 2013 (SWG 2016), confirmed what local Inuit and some biologists had been saying for years: contrary to the assertions of PBSG scientists[35] , BB polar bear numbers have not declined since 1997 due to suspected over-hunting.[36] BB bear abundance in 2013 was found to be 'considerably larger' than the previous estimate but the authors assert that differences in sampling design preclude direct comparison between the two.[37] Still, the polar bear subpopulation estimate at 2013 for BB was 2,826±767 (95% CI = 2,059–3,593), a 36% increase over 1997 (2,074; 95% CI = 1,553–2,595). While all other metrics of life history and habitat were subject to statistical significance testing, the abundance estimate was not, because of the claimed methodological issues (a position refuted by Mitchell Taylor, the author of the 1997 report).[38]

Davis Strait – Seasonal

There have been many post-hoc adjustments made to the first population estimate for Davis Strait (DS) bears, which include those that visit Newfoundland and southern Labrador in the spring. The original count (completed in the late 1970s), generated a figure of 726 bears, later adjusted to 900.[39] That estimate was subjectively increased by the PBSG to 1,400 bears in 1993,

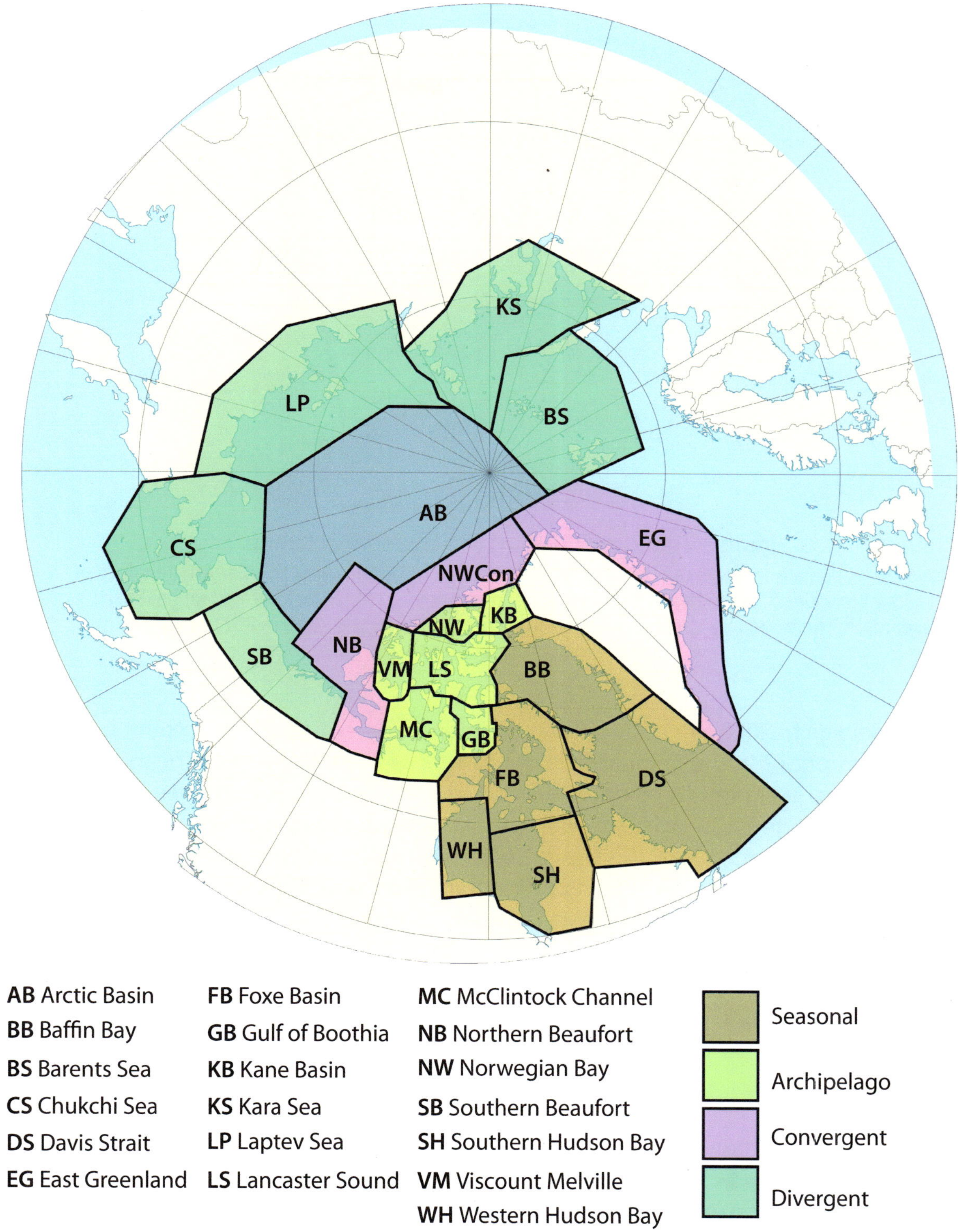

Figure 2: The four Arctic sea ice ecoregions.

The Arctic Basin (AB) is not considered to be a sea ice ecoregion. The Convergent region 'NWCon' (also known as 'Queen Elizabeth – Convergent') is not a recognized polar bear subpopulation.

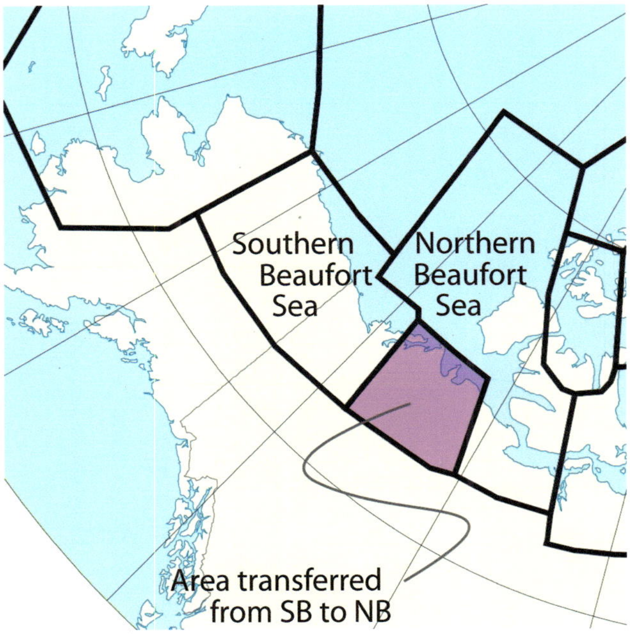

Figure 3: Proposed Beaufort Sea boundary change.

The change virtually restores the boundaries of what used to be known as the 'Eastern Beaufort' in older literature,[31] which has been maintained in recent years for polar bear and ringed seal research.[32] The boundary change was provisionally adopted by the PBSG in 2017 but so far, no official maps have been altered.

and increased again in 2000 to 1,650 (without additional field surveys)[40] to account for certain biases and assumptions in the original estimate as well as more sightings of bears and an increase in their harp seal prey. A comprehensive survey completed in 2007 generated a new estimate of 2,158 (range 1,833–2,542),[41] a substantial increase over the previous estimates. The density of Davis Strait bears in 2007 (5.1 bears/1,000 km^2) was found to be higher than other seasonal sea-ice subpopulations such as Hudson Bay. Karyn Rode and colleagues recorded a slight decline in body condition of DS polar bears between 1977 and 2010 but there was no indication this had affected survival or reproduction.[42] By 2012, the harp seal population had grown even further,[43] providing the potential for a further increase in polar bear numbers and this is probably reflected in the 2014 Environment Canada status assessment as 'likely increasing'.[44]

Foxe Basin – Seasonal

The first survey of Foxe Basin (FB) bears, conducted in 1994, generated an estimate of 2,197 bears (1,677–2,717),[45] but in 2004 this was adjusted by the PBSG to 2,300 bears (1,780–2,820). An aerial survey in 2009–2010 generated an estimate of ~2580 bears, the first aerial surveys performed in Canada after mark-recapture studies were effectively banned by the Nunavut government.[46] While the two methods (aerial survey and mark-recapture) are not directly comparable, the population was considered stable by Environment Canada and the PBSG in 2014, and there is no recent evidence to conclude this is not still the case.[47]

Western Hudson Bay – Seasonal

The first comprehensive survey of Western Hudson Bay (WH) for the period 1978–1992[48] generated a population estimate of 1,000±51, which was adjusted by the PBSG in 1993 to 1,200[49] to account for areas not surveyed. Regehr and colleagues estimated the abundance in 2004 as 935 (range 794–1,076),[50] a statistically significant decline of 22% from the 1,194 bears counted in the same core area in 1987 (range 1,020–1,368). This result was used as persuasive evidence that polar bears should be listed as 'vulnerable' and 'threatened' (by the IUCN Red List in 2006, and the US Fish & Wildlife Service in 2008, respectively). A mark-recapture study in 2011 of the core region only[51] generated an estimate of 806 (653–984) but an aerial survey the same year that encompassed the entire subpopulation area generated an estimate of 1,030 (range 754–1,406),[52] which is still the estimate used by the PBSG in 2017.[53] An aerial survey in 2016[54] used a somewhat different method from the 2011 survey and generated an estimate of 842 (range 562–1,121).

Because the 2011 and 2016 Western Hudson Bay aerial surveys used somewhat different methods, the only population-size numbers that can be compared are these figures from the same area: 2011 (estimate 949, range 618–1280) and 2016 (estimate 842, range 562–1121). The slight decline apparent over five years (11%) was not statistically significant.[55] When differences in methodology and assumptions are taken into account, there is no evidence to suggest the estimate for 2016 of 842 bears is different from the 2011 estimate of 1030 bears, neither of which is statistically different from the estimate of 935 calculated in 2004, which means there has been no statistically significant decline in abundance since 2004.

Southern Hudson Bay – Seasonal

The first population size assessment for Southern Hudson Bay (SH) was made during 1984–1986, and generated an estimate of 763±323 bears.[56] Some adjustments, re-analyses and new surveys indicated that by 2005 the subpopulation had been stable since the mid-1980s.[57] A subsequent aerial survey in 2011–2012 generated an estimate of 943 bears (range 658–1350).[58] Results of the most recent survey, conducted in 2016, have not yet been released.

Barents Sea – Divergent

The first count of Barents Sea (BS) polar bears was undertaken in August 2014 using a combination of mark-recapture and aerial survey over both Norwegian and Russian territories. This survey generated an initial estimate of 2,997[59] , which was later amended to 2,650 (range 1900–3600) for the entire region.[60] Researchers found 2.87 times as many bears in the Russian sector of the Barents Sea as in the Norwegian sector in 2004.[61] In August 2015, a planned recount of the subpopulation had to be restricted to the Norwegian sector because Russian authorities refused to issue the necessary permits.

The Svalbard area count was initially reported to have increased by 42% over the count performed in 2004, with most bears found to be in good or excellent condition. The published peer-reviewed paper confirms that a 42% increase in abundance indeed occurred (685 bears in 2004 to 973 bears in 2015),[62] but due to the large uncertainty (broad error ranges) in the estimates involved, that 42% increase was not statistically significant. This point was strongly emphasized by the PBSG in their recent status update for the Barents Sea: Because of the overlapping confidence intervals, it cannot be concluded that the BS subpopulation has grown, but it can be considered stable.

The authors of the Svalbard survey had this to say:

> There is no evidence that the fast reduction of sea-ice habitat in the area has yet led to a reduction in population size. The carrying capacity is likely reduced significantly, but recovery from earlier depletion up to 1973 may still be ongoing.[63]

The same authors also concluded that only a few hundred bears use Svalbard routinely as a denning area or summer refuge and that most individuals seen around the area live in the pack ice offshore, confirming their previous finding that most Barents Sea polar bears live in the Russian sector of the region.

Zoologist Susan Crockford pointed out in 2017[64] that if the results of the 2015 survey were extrapolated to the entire region using the ratio for the Russian and Norwegian sectors from the 2004 survey, the 2015 population size for the Barents Sea would likely be about 3,749 (an increase of about 1,109 bears). This extrapolation might not be correct or statistically significant but accounts for the high probability that the polar bear population in the Russian sector increased between 2004 and 2015 by at least as much as the Norwegian sector (and perhaps by even more, because sea-ice conditions there have been less seasonally volatile).[65] The 2017 PBSG did not use the results of the 2015 survey to update BS numbers from 2004 because the new Svalbard count was not considered statistically significant, which means the official population size remains at 2,650 (range 1900–3600).[66]

Kara Sea – Divergent

A first-ever Kara Sea (KS) population estimate completed in late 2014[67] potentially adds another 3200 or so bears to the global total. This estimate (range 2,700–3,500), derived by Russian biologists from ship counts, was added to the official global count published in 2015 by the IUCN Red List.[68] An earlier 'ballpark' estimate was about 2,000 individuals, suggesting an increase may have taken place. However, the PBSG in 2017 still lists Kara Sea as 'unknown' and does not mention the 2014 Russian estimate.[69]

Laptev Sea – Divergent

The Laptev Sea (LS) was given a population size of about 1,000 (range 800–1000) based on den counts in the 1960s to 1980s.[70] The PBSG included this estimate in its 2005 assessment,[71] but this status was changed to 'data deficient' in 2013 and 'unknown' in 2014,[72] due to the estimate being out of date. 'Unknown' was the PBSG status in 2017.[73] However, the 2015 IUCN Red List assessment required population size numbers be used in any models for projecting future status, so the estimate of 1,000 was considered accurate enough when there was nothing better.[74]

Chukchi Sea – Divergent

There has been no comprehensive survey for the Chukchi Sea (CS), but a previous estimate of 2,000 was the number used for the 2015 Red List assessment,[75] although the current PBSG table lists no estimate and gives the status as 'unknown.' An estimate of 3,000–5,000 based on den counts and estimated numbers of females in the population by Russian researchers became 2,000–5,000 in the 1993 PBSG report and 2,000 in the 2005 report.[76] Considered 'declining' by the PBSG in 2009 based on current and projected sea ice losses,[77] that changed to 'data deficient' in 2013[78] and 'unknown' in 2014–2017[79] .

However, a strong argument can be made that this subpopulation qualifies as 'stable'. Evidence collected up to 2016 suggests that bears are in good condition and reproducing well.[80] Research conducted from 2008–2011 showed that CS polar bears were doing better than they were in the 1980s, and body condition was better than any other subpopulation except the bears of Foxe Basin (northern Hudson Bay), who were doing exceptionally well. Bears that spend the summer on Wrangel Island, the region's main terrestrial denning area, had increased dramatically, from about 200–300 in 2012 and 2013 to 589 in the fall of 2017,[81] although about 550–600 were counted in 2007.[82] Eric Regehr was quoted on 23 November 2017 as saying the subpopulation 'appears to be productive and healthy'.[83]

Southern Beaufort Sea – Divergent

The first Southern Beaufort Sea (SB) survey in 1986 generated an estimate of about 1,800 individuals. It attempted to take into account known movements of bears to and from the Chukchi Sea to the west and the Northern Beaufort Sea to the east.[84] Such movements were what prompted the change in the SB/NB boundary in 2014; a similar change in the western boundary (near Barrow, Alaska) has been discussed but not implemented.[85] Mark-recapture studies in 2001–2006 generated a statistically insignificant decline to about 1,526 (range 1,211–1,841), which was subsequently blamed on reduced summer ice,[86] although it was clear that a series of thick spring sea ice episodes from 2004–2006, as severe as had occurred in 1974–1976, was ultimately responsible for the poor survival of cubs, reduced body condition of adults and subadults, increased spring fasting, and the reduced abundance of ringed seals.[87]

Additional survey data from 2007–2010, analysed using a totally new method, showed that survival picked up in 2007 (just as summer sea ice hit a record low) and increased through 2009, resulting in a revised estimate of 907 (range 548–1270) in 2010, a statistically significant decline of roughly 25–50% (often cited as '40%') over the 1980s count.[88] The PBSG point out in 2017 that the latest survey may not have sampled the entire geographic range adequately, and that this may have negatively skewed the 2010 population estimate:[89] they did not, however, make an adjustment to the population estimate as they had previously done when such problems with estimates later became evident (e.g. Davis Strait).

Northern Beaufort Sea – Convergent

The last population count for the Northern Beaufort Sea (NB) was made in 2006, generating an estimate of 980 (range 825–1135).[90] Compared to earlier assessments, the population appeared at that time to have been relatively stable over the previous three decades. The boundary with Southern Beaufort has been moved east to near Tuktoyaktuk for Canadian management purposes, a change provisionally accepted by the IUCN PBSG in 2017,[91] but this change is not yet incorporated into any polar bear range maps.

East Greenland – Convergent

Although there has been no comprehensive survey of the East Greenland (EG) subpopulation, as recently as 2001 the PBSG gave a rough estimate for this subpopulation of 2,000 bears (in part based on harvest records that indicated a fairly substantial population must exist).[92] However, in 2013 they credited the region with only about 650 bears, with no reason given for the change in opinion.[93] By 2014, this subpopulation was simply said to be 'very low'.[94] Traditional ecological knowledge (TEK) from hunters suggested an increase in numbers by 2011, but this has not yet been substantiated. The population was officially considered 'unknown' in 2017,[95] but the original estimate of 2,000 bears was considered adequate for the 2015 IUCN Red List assessment.[96]

Arctic Basin – a designated subpopulation but not an ecoregion

In the original classification of the sea-ice ecoregions, a narrow portion of the Arctic Basin (AB) north of Greenland and Ellesmere Island were called 'Queen Elizabeth – Convergent' and the later, 'Northwest – Convergent' (NWCon; Figure 2), but that now seems to have been abandoned, probably because it is not a distinct subpopulation region for polar bears.[97] The PBSG considers the Arctic Basin in general to be a 'catch-all' region because it contains bears moving between regions and those from peripheral seas (such as the Southern Beaufort and Barents[98]) who use it as a summer refuge during the ice-free season.[99] Both single bears and family groups have been seen feeding on ringed seals during the summer, and both ringed seals and their fish prey have been documented as being present.[100] The Arctic Basin is given a population size estimate of zero[101] but there is some evidence that the productivity in some areas of this region is higher than previously assumed[102] and it is thus possible that a small number of polar bears may live there year-round.

Kane Basin – Archipelago

A 2013 survey of Kane Basin (KB) polar bears confirmed what local Inuit and some biologists have been saying for years: that contrary to the assertions of PBSG scientists, KB polar bear numbers have not been declining.[103] Until recently, the KB polar bear subpopulation, located between north-west Greenland and Ellesmere Island, was assessed with confidence by the PBSG to be declining due to suspected over-hunting. In 2014, Environment Canada's assessments were 'data deficient' for the area.[104] The 2013 survey generated an estimate of 357 (range 221–493), a 59% increase over the 1997 estimate (the latter recalculated in 2016 as 224 (range 145–303) from the original estimate of 164 (range 94–234) in 1994–1997), indicating a 'stable to increasing' population.[105] However, the authors expressed concerns with sampling methodology and differences in the areas surveyed and suggested 'some caution in interpretation of population growth' was necessary.[106] While all other metrics of life history and habitat were subject to statistical significance testing, the authors do not state conclusively whether the 59% increase was statistically significant or not but the PBSG assessment for 2017 concluded that the population has indeed increased.[107]

McClintock Channel – Archipelago

The first population size estimate generated for McClintock Channel (MC) was about 900 bears in the mid-1970s and a mark-recapture study in 2000 generated an estimate of 284±59 bears, a significant decline blamed on over-hunting.[108] Hunting was subsequently halted then resumed at a much reduced level, after which the population was presumed to be increasing. A three-year mark-recapture study began in 2014[109] and results may be available in 2018.

Viscount Melville – Archipelago

The first survey to determine the population size of the Viscount Melville (VM) subpopulation was completed in 1992 and generated an estimate of 161±40.[110] This estimate is now 25 years old and a new one is in progress: results may be released in 2018.[111]

Gulf of Boothia – Archipelago

The Gulf of Boothia (GB) is in the middle of the Canadian Arctic. In terms of geographic area, it is one of the smallest of all 19 subpopulations worldwide: at only 170,000 km^2, only the Norwegian Bay and Kane Basin regions are smaller, at 150,000 and 155,000 km^2 respectively.[112] The first population survey was done in 1986 and generated an estimate of about 900 bears. This was updated in 2000, with an estimate of 1,592±361 bears, a significant increase.[113] The new density was calculated as 18.3 bears per 1000 km^2, well above the 5.1 bears per 1000 km^2 found in Davis Strait, 1.9 bears per 1000 km^2 in McClintock Channel, and 6.5 bears per 1000 km^2 found in the Northern Beaufort Sea.[114] A new estimate for the area is pending as of 2017.[115]

Lancaster Sound – Archipelago

The Lancaster Sound (LS) subpopulation, in the middle of the Canadian Arctic archipelago, has one of the highest populations of polar bears anywhere, although it is one of the smaller regions: only the Barents Sea and Foxe Basin have higher estimated population sizes. The latest population surveys in LS were conducted from 1995 to 1997, and in 1998 an estimate of 2,541±391 bears was generated, a significant increase over the previous estimate (1977) of 1,675 bears.[116] The eastern portion of LS is generally clear of ice by late summer (hence the Northwest Passage) but the western third of the region not only retains pack ice later in the season but some multiyear ice remains throughout the year. The proximity of LS to Baffin Bay and the eastern Northwest Passage undoubtedly exposed polar bears there to hunting by European whalers during the 1800s and early 1900s,[117] but the population appears to have recovered since then. In 2017, the PBSG considered the population to be stable.[118]

Norwegian Bay – Archipelago

The last population count for Norwegian Bay (NB) was done in 1993–1997 and generated an estimate of 203±44.[119] The figure is thus now well out of date.[120] Several genetic studies suggest this may be a genetically distinct subpopulation.[121] Norwegian Bay is either part of, or adjacent to, what has been called the 'Last Ice': a refugium of sea ice over shallow continental shelf waters expected to remain even if summer sea ice drops to near-zero levels (<1 million km^2), depending on the model used.[122]

4 Population trends

In 2014, Environment Canada published a global polar bear population status and trend map that showed two subpopulations were likely declining, two likely increasing, and six stable or likely stable, with nine being data deficient for these purposes.[123] Although this map has not yet been updated, by 2017 additional results of population surveys and other studies would likely change these classification totals to:

- four 'likely increasing' (KS, KB, DS, MC)
- nine 'stable' or 'likely stable'*
- one 'likely declining'
- five data deficient.

However, instead of using this knowledge to conclude that bears in unstudied or data-deficient regions are probably stable or increasing as well, IUCN PBSG members illogically suggest bears in those regions could be in trouble without anyone knowing. Figure 4 shows those regions as 'presumed stable or increasing.'

Only the Southern Beaufort (SB, Divergent sea ice ecoregion) registered a statistically significant decline at its last population count, but this region has special circumstances that make it an outlier: what happens in SB is not representative of any other polar bear subpopulation. A decline in numbers occurred during a period when summer sea ice had declined markedly (2001–2010) but the proximate cause of the population decline was a series of thick spring ice events in 2004–2006 that drove seals out of the region.[125] This situation is unique to the Southern Beaufort, although something similar happens on a less regular basis off Greenland and in Hudson Bay due to thick ice and/or changes in snow depth over ice.[126]

While the PBSG listed population trends in their population status table in 2014 (when they considered four populations to be declining), by early 2017 they no longer include this hitherto important metric.[127] Since no rationale for the removal of the trends assessment has been given, the PBSG appears to be exhibiting a reluctance to declare that stable polar bear populations are now the norm across the Arctic (Figure 4).

---

* With the Chukchi Sea trend based on remarks by US Fish and Wildlife biologists Eric Regehr to *The Daily Mail* (23 November 2017) that this subpopulation 'appears to be productive and healthy'.[124]

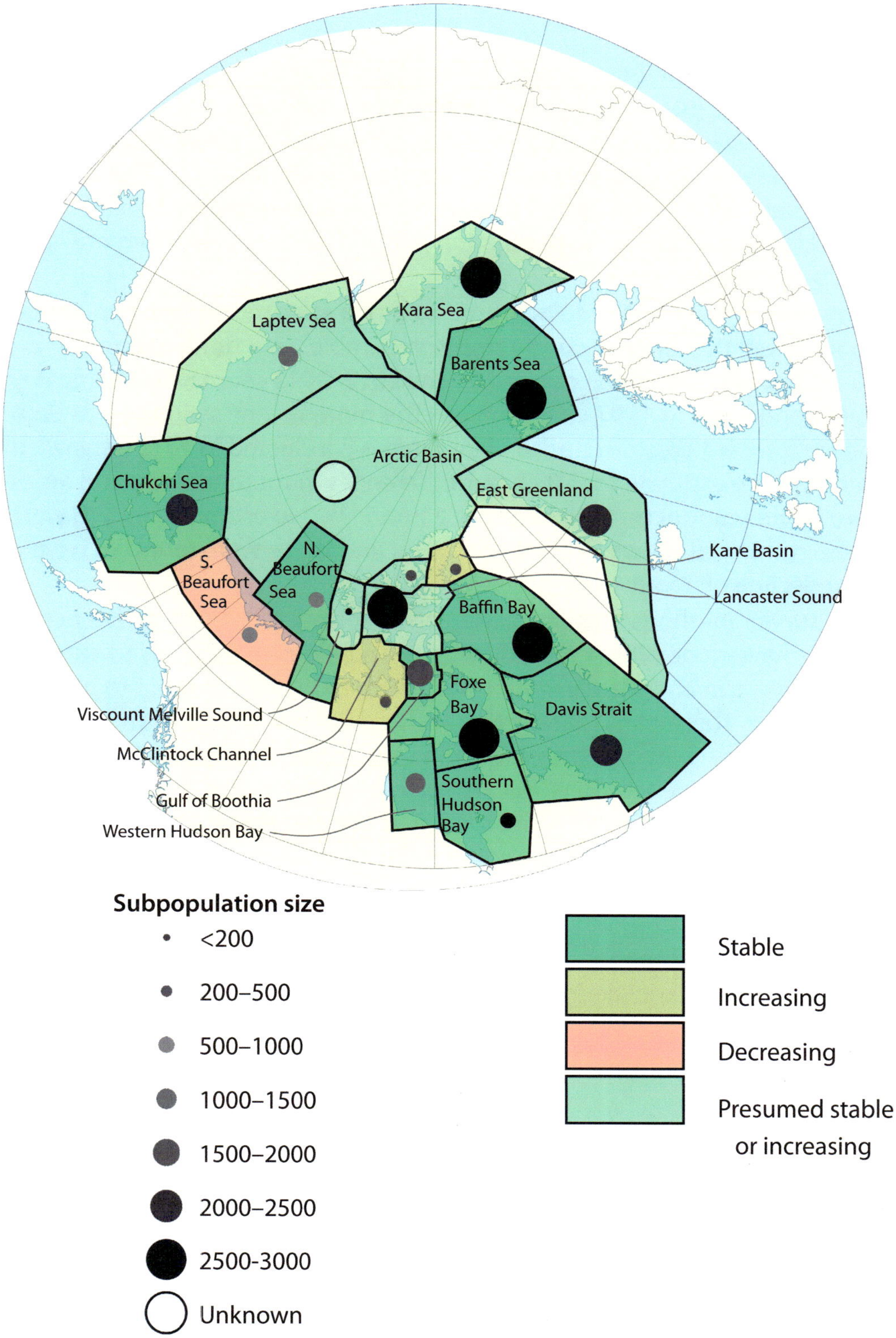

Figure 4: Trends in polar bear subpopulations at 2017.

Former 'data deficient' regions are marked 'likely stable or increasing' to reflect current research on studied populations. Modified from WWF population size and trends map for 2017.[128]

The problem of statistical confidence

Virtually all recent population-size estimates for polar bear subpopulations have such wide margins of error (statistical confidence intervals) that even quite large changes in size are unlikely to be statistically significant. For example, in its most recent area population count in 2015, Svalbard saw a population increase of 42%, but this was not statistically significant.[129] The authors, Jon Aars and colleagues, could conclude only that recent large declines in sea ice habitat in the Svalbard area had not yet led to a reduction in population size and that recovery from previous overhunting might still be ongoing.

In addition, differences in survey methodology used to arrive at particular subpopulation estimates have led to numbers that are not considered comparable, so a trend cannot be established. Such problems have recently been claimed for the 2012–2013 estimate for Baffin Bay bears compared to one conducted in 1997, even though the more recent estimate was 36% larger.[130] Similarly, the most recent Western Hudson Bay surveys conducted in 2016 generated estimates 33% smaller than the estimate for 1987, but differences in methodology and areas surveyed mean the two figures cannot be used to derive a trend.[131] Two estimates for WH were generated that could be compared between 2004 and 2016, and these were found to have a statistically insignificant decline of about 11%.[132]

In short, changes in survey methods and/or mathematical formulas used to derive population estimates over time have generally increased statistical confidence intervals to such an extent that a decline or increase in abundance would likely need to be 50% or more to be considered a real and valid change. This means that the ESA and Red List definitions of 'threatened' with or 'vulnerable' to extinction – based as they are on the likelihood of a population decline of 30% or more over the next three generations[133] – are using a mathematical threshold that is very likely statistically invalid for polar bears.[134] The IUCN Red List assessment for 2015 apparently dealt with this issue by concluding that there was a reduced probability (only 70%) that a decline of 30% or more would occur by 2050.[135]

I have dealt with this issue in this report by replacing old population sizes with new ones generated since the 2015 Red List assessment was published (for Barents Sea, Baffin Bay, Kane Basin, Western Hudson Bay)[136] but acknowledge that the margin of error remains large and note the apparent increase in global population size is likely not statistically significant. The rationale for this approach is to emphasize that the anticipated decline in global numbers since 2005 has not taken place.

5 Habitat status

Global sea ice

Summer sea ice (at September) has declined markedly since 1979, especially since 2007, but winter ice levels (at March) have declined very little (Figure 5). There has been no research done on what effects, if any, the slight decline in winter ice extent has had on polar bears overall, but a cursory examination suggests that since 1979 there has been enough sea ice in winter to meet the needs of polar bears and their prey. As far as is known, record low extents of sea ice in March 2015 and 2017[137] had no impact on polar bear health or survival (Figure 6a). For example, adult male bears captured around Svalbard, Norway, showed no statistically significant change in condition during the spring of 2017 or 2015, compared to those captured from 1993 to 2017.[138]

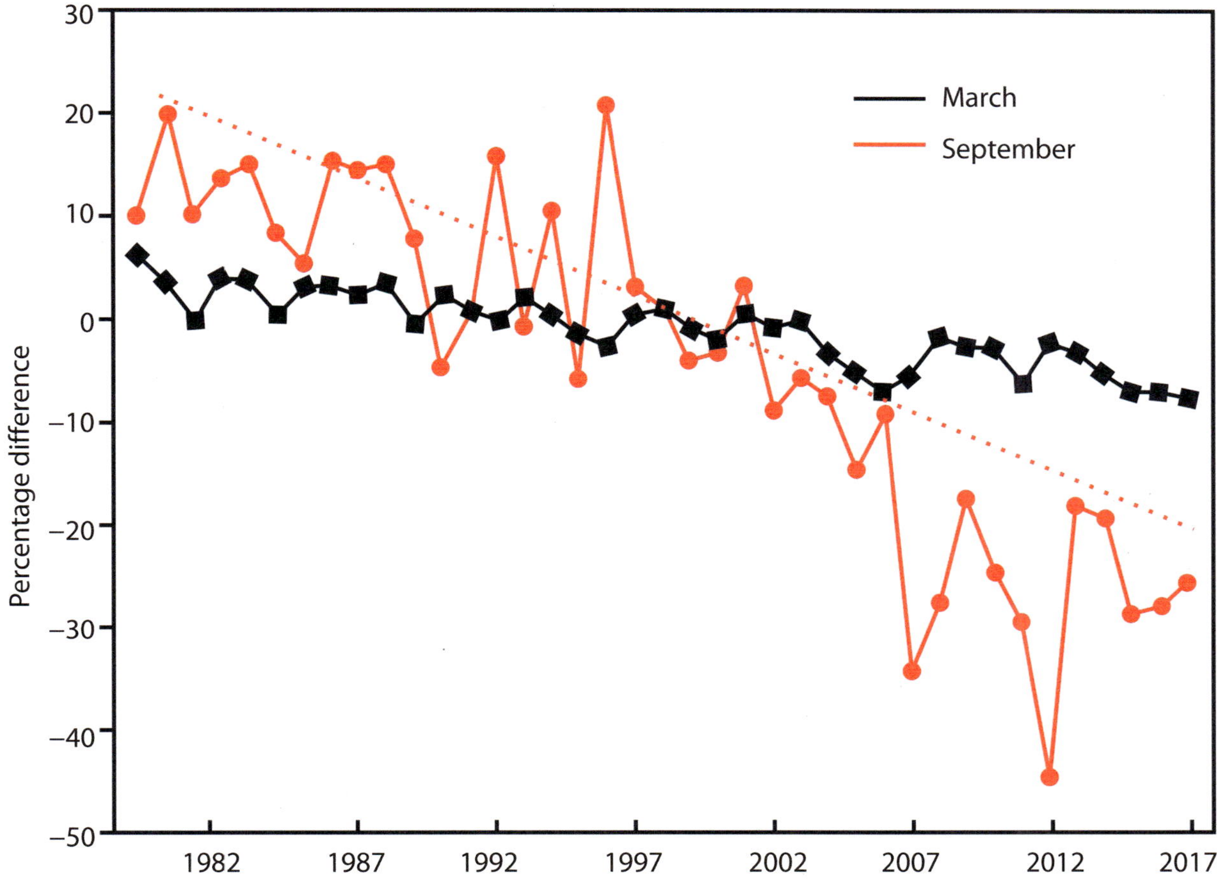

Figure 5: Sea ice extents.

Minimum extent for September (summer, red) and maximum extent for March (winter, black), 1979–2017, from NOAA's Arctic Report Card 2017.[139] A recent paper tracking sea ice levels back to 1850 shows a similar pattern.[140]

The most pessimistic predictions of March sea-ice extent at the end of the 21st century is about 12.0 million km^2,[141] equal to the average extent of ice for May 2016 (Figure 6b). Polar bears and their prey could survive without a precipitous decline in population size if March sea ice dropped this low, even before 2100, because there would be enough ice in all regions where these animals reside to meet their minimum spring requirements.

Sea ice extent in June has declined, on average, from just over 12 million km^2 in the 1980s to just over 11 million km^2 from 2004–2017.[142] By late May to early June, there is therefore lots of sea ice throughout the Arctic to act as a feeding platform for polar bears (Figure 6c,d). However, the young seals that form the bulk of polar bear diets in spring take to the water to feed and

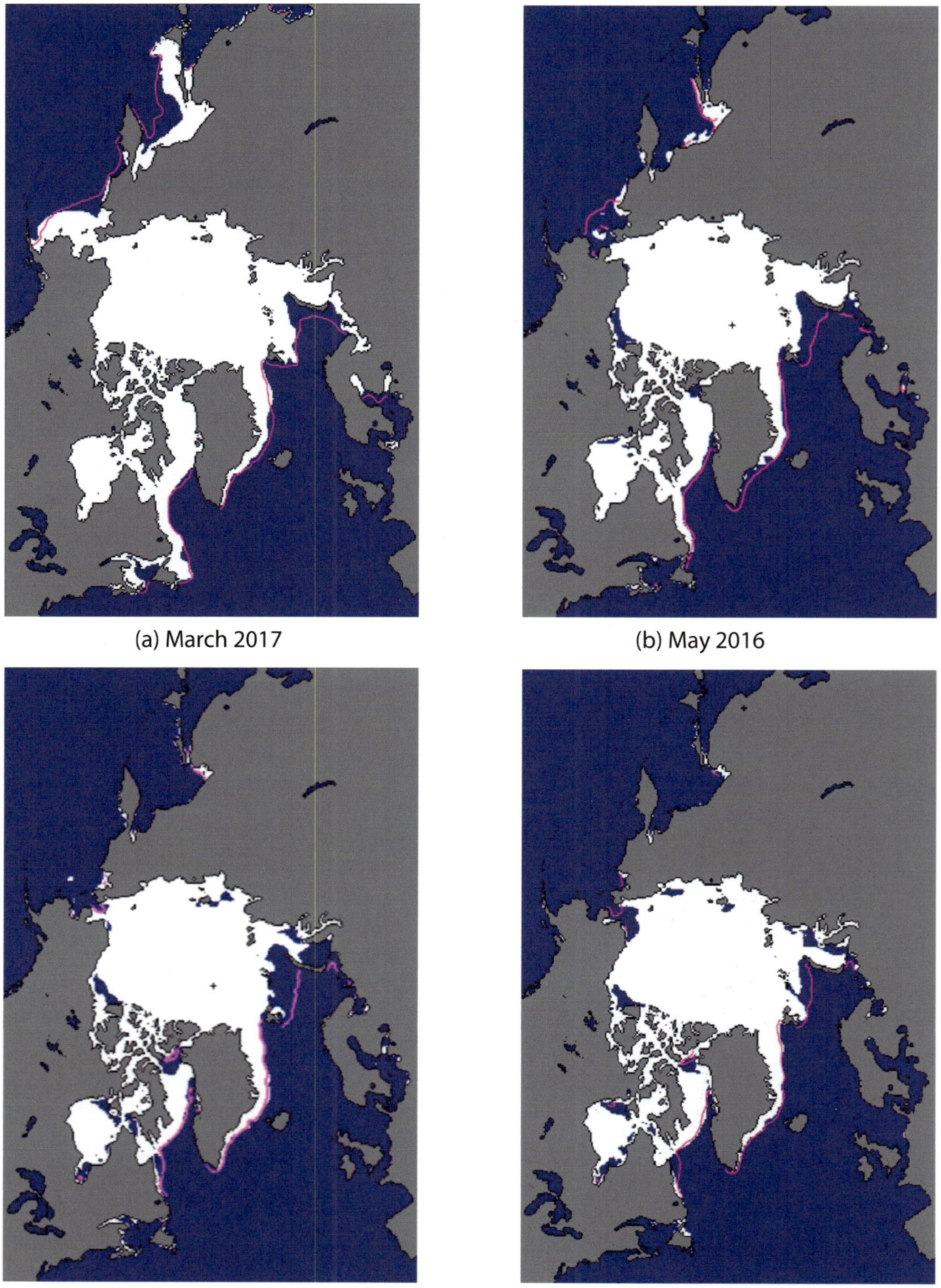

(a) March 2017

(b) May 2016

(c) June 2012

(d) June 2017

Figure 6: Average sea ice extents.
The pink lines indicate the median extent. Courtesy US National Snow and Ice Data Center.

are no longer available on the ice, leaving only predator-savvy adults and subadults hauled out on the ice as potential prey.[143] This means few seals are actually caught and consumed by polar bears after about mid-June in seasonal and divergent sea ice ecoregions, or by mid-July in convergent and archipelago regions (see Section 6).

Sea-ice thickness has declined in some regions of the Arctic, but by and large this has been a net benefit for polar bears and their prey whose preferred habitat is first-year ice less than two metres thick.[144] For example, sea ice in Kane Basin, west of Northern Greenland, was predominantly multi-year ice even in summer during the 1980s, and this poor seal habitat supported few polar bears. But now that the ice is mostly seasonal first-year ice, the population of bears has grown remarkably.[145]

In contrast, a just-released report about Southern Beaufort Sea bears having difficulty finding prey in 2014–2016[146] indicates that the thick ice events that have impacted the region every ten years or so since the 1960s have continued despite reduced summer sea ice, although Pagano and colleagues do not draw that conclusion.[147] The scientific literature has many papers and reports that show what past episodes of thick spring sea ice have done to polar bears that live in the Southern Beaufort Sea.[148] There is also evidence the phenomenon occurred again in 2014–2016, right on schedule, ten years after the 2004–2006 episodes, although researchers and the media[149] blamed the effects on reduced summer sea ice.[150] The devastating effects that heavy ice cover has had on polar bears in the Beaufort Sea has been documented for 1974–1976, 1984–1986, and 2004–2006, with similar events inferred from anecdotal information for 1964 and 1992.[151] Susan Crockford argued a few years ago that Arctic sea ice is not the stable habitat that polar bear experts currently assume,[152] which means that population numbers in some regions will vary naturally in response. This was a conclusion reached by Ian Stirling in 1982, and warrants repeating here:

> Until recently, management of marine mammals in the Canadian Arctic, to the extent that they are managed at all, seems to have been based on the assumption that ecological conditions show little variability. Thus, once populations are counted or quotas are established, little change in population management takes place for long periods. The results of this study have clearly shown that ice conditions in the eastern Beaufort Sea can be highly variable, can influence other ecological parameters, and can cause changes in the distribution and abundance of ringed and bearded seals. We expect that similar variability will be documented in other areas of the Arctic when comparable studies have been completed. What this means in terms of environmental assessment is that, because conditions are so variable, the consequences of possible man-made detrimental effects will vary depending on the status of the seal populations at the time.[153]

While polar bear specialists have for years insisted that polar bears prefer sea ice of 50% or more over continental shelves regardless of season, recent research has shown bears utilize sea ice during the melt season that is well below this threshold. In the Southern Beaufort Sea, as well as in Western Hudson Bay, bears were found to use ice of 0–20% concentration and in some cases SB bears were tracked to areas registered by satellites as open water.[154]

Sea ice varies between seasons, of course, but it is often highly variable from year to year within a sea ice ecoregion and across the Arctic as a whole. Over longer periods of time (decades, centuries, millennia), Arctic sea ice has also been quite variable, at times more extensive than today and at others, less extensive.[155] For example, the abrupt decline in sea ice extent that came at the end of the Younger Dryas cold period (ca. 10,000–12,500 years ago), especially in the Eastern Arctic, where ice had extended into the North and Baltic seas in summer, meant an abrupt contraction of range: most of the polar bear fossils of the Younger Dryas come from Denmark, southern Sweden and southern and western Norway.[156] The Younger Dryas ended abruptly, perhaps as a result of a comet strike,[157] over a 40-year period; the change took place in a series of steps of about five years' duration each.[158] Polar bears and their prey species (such as ringed and bearded seals, walrus, beluga, and narwhal) have survived these and other changes with no apparent negative effects.[159] Their inherent flexibility in dealing with changing ice conditions means that evolutionary adaptation, as it is commonly defined, has not usually been necessary.

Sea ice loss by subpopulation

Eric Regehr and colleagues[160] provide details of the amount of sea-ice loss (number of days with ice cover of >15% concentration) per year for the period 1979–2014 per polar bear subpopulation. This metric varied from a high of 4.11 days per year in the Barents Sea to a low of 0.68 in the southernmost region, Southern Hudson Bay. Most subpopulations have lost about one day per year since 1979, although a few have lost somewhat more or less.[161] Surprisingly, as Table 1 shows, despite having the greatest loss of ice since 1979, polar bear numbers in the Barents Sea in 2015 have grown since 2004, and bear numbers in Southern Hudson Bay, with the least amount of ice loss, have remained stable since the 1980s.[162]

Freeze-up and breakup date changes

Freeze-up for Hudson Bay came as early in 2017 as it did in the 1980s. This allowed most WH and SH polar bears to resume seal hunting four weeks earlier than last year, when freeze-up was quite late. WH bears leave the shore within about two days of sea ice concentration reaching 10% along the shore, although SH bears leave when it reaches about 5%:[163] in other words, the bears leave shore as soon as they possibly can. This year, there was enough ice by 8 November for many bears to leave shore and by 10 November most bears were on their way. According to data for 1979–2015, in the 1980s the mean date that bears left the ice at freeze-up (10% sea ice coverage in WH) was 16 November±5 days, while in recent years (2004–2008) the mean date of leaving was 24 November±8 days, a difference of 8 days.[164] This also means that a freeze-up date of 10–12 November (day 314–316) for 2017 was one of the earliest freeze-up dates since 1979 (the earliest being 6 November, day 310).

Despite the overall drop in ice-covered days since 1979 there has been no statistically significant change in either breakup or freeze-up dates for WH since the mid-1990s.[165] Most of the change, an increase in the ice-free period of about three weeks, came about 1998. The ice-free season has increased in SH by about 30 days, but, as for WH, most of that change came in the late 1990s, with much yearly variation in breakup and freeze-up dates since then.[166]

Table 1: Sea ice loss per subpopulation.

Subpopulation by sea ice ecoregion	Days lost per year (1979–2014)
Seasonal	
Baffin Bay	1.27
Davis Strait	1.71
Foxe Basin	1.15
Western Hudson Bay	0.86
Southern Hudson Bay	**0.68**
Divergent	
Barents Sea	**4.11**
Kara Sea	1.70
Laptev Sea	1.35
Chukchi Sea	0.90
Southern Beaufort Sea	1.75
Convergent	
East Greenland	1.07
Northern Beaufort Sea	0.93
Archipelago	
Kane Basin	1.44
McClintock Channel	1.12
Viscount Melville	1.26
Gulf of Bothia	1.88
Lancaster Sound	1.08
Norwegian Bay	0.73

Change in number of days with ice cover of >15% concentration per year. From Regehr and colleagues (2016). Lowest and highest values in bold.

6 Prey base

Ringed and bearded seals are the primary prey of polar bears worldwide. In some regions, other seal species make up varying proportions of the diet: harp seals for Davis Strait and East Greenland bears, and ribbon seals for Chukchi Sea bears, for example. Walrus, beluga, and narwhal make up a small proportion of the diet in some areas. Arctic seals have their pups and mate in the spring on the sea ice. Pups gain weight rapidly and are weaned after a short nursing period, after which mating occurs. Most ringed and bearded seals, as well as several less common species, are born from mid-March to mid-April or a bit later, depending on the location; harp seals are born earlier (February to mid-March) in less consolidated pack ice than the others.[167] Newborns are preyed upon by polar bears soon after they are born.

Seal pups are about double their birth weight or heavier by the time they are weaned, when they may be 50% fat by weight. Male seals hang around the mother–pup pair and mate with

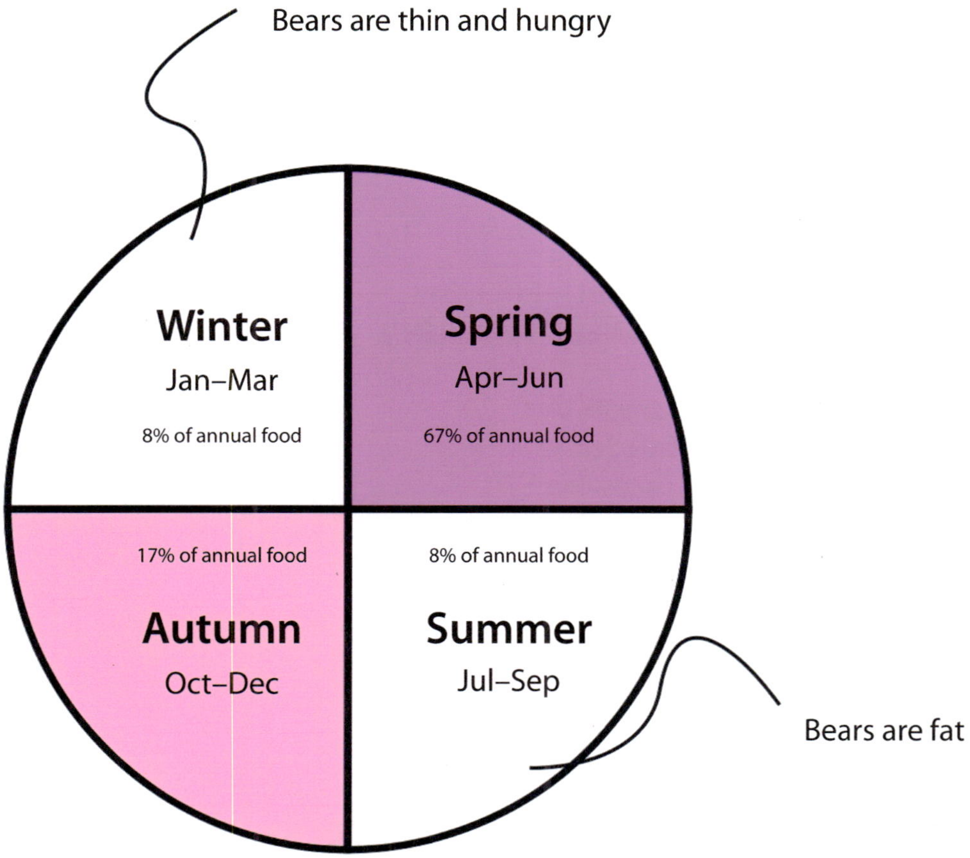

Figure 7: Polar bear feeding activity by season.
The most intensive feeding time is spring, followed by fall. Although some individuals have trouble eating enough in the spring due to inexperience, competition, old age, injury or disease, polar bears are usually hungriest in late winter, not summer as some people believe.[168] Based on data from the polar bear literature, seasons as defined by Pilfold and colleagues in 2015.[169]

the female around the time the female stops nursing. Newly-weaned seals are an important food source for fat-craving polar bears in early spring for the three or four weeks that they are available. After that time, young seals take to the water to feed and are largely unavailable to polar bears. In some areas, polar bears can kill up to 44% of newborn seal pups each spring if conditions are right.[170]

Both polar bears and Arctic seals need ice in the spring (April–June): polar bears for hunting, seals for nursing pups and mating. Polar bears consume two-thirds of their yearly food supply in the spring (Figure 7). Even though summer sea ice has routinely declined to less than 5 million km^2 in recent years, there has still been plenty of ice remaining to act as a hunting platform for polar bears until the middle or the end of June or later, depending on the location. From May to July, adult and subadult seals of all varieties haul out on sea ice while they moult, but they are harder for polar bears to catch than youngsters because they are predator-savvy.

While it is true that some Arctic ringed seals give birth in stable shorefast ice close to shore,[171] many others give birth well offshore in thick pack ice where polar bears also live and hunt in the spring.[172] Although not often mentioned, there is documented evidence of pack-ice-breeding ringed seals in the Bering Sea, Sea of Okhotsk, Chukchi Sea, Davis Strait, and the Barents Sea. This finding is supported by genetic evidence.[173] The presence of breeding ringed seals in the pack ice suggests strongly that polar bear spring hunting habitat includes all Arctic sea ice of suitable thickness over continental shelf waters, not just shorelines and fjords.[174]

While ringed seals and bearded seals were both listed as 'threatened' under the US Endangered Species Act in 2012, there is no evidence that either species has declined in number or registered any other negative impact due to reduced summer sea ice.[175] The 'threatened' status is based exclusively on the presumption that future harm will result from further reductions in summer sea ice.[176] However, no other Arctic nation has taken this conservation step for ringed and bearded seals, and neither has the IUCN Red List, which lists both as 'Least Concern'.[177]

Harp seals are an important alternate prey for polar bears in Davis Strait, Baffin Bay, East Greenland, and the Barents Sea. By 2015, there were an estimated 7.4 million harp seals in Atlantic Canada (range 6.5–8.3 million), an exponential increase over the early 1980s when perhaps only half a million remained.[178] Relatively few harp seals give birth and breed in the Gulf of St. Lawrence (where there are no polar bears) but in some years they have suffered increased mortality due to reduced spring ice conditions, something that has happened more often in recent years.[179] However, the seals that whelp off Labrador and Newfoundland (where they are the main prey of polar bears) appear to have been less vulnerable to such changes, in the past and in recent years. From 1950–1990, there were poor ice conditions in the Gulf about one year in every ten, but, as in the past, most animals today likely move to the 'front' ice off Newfoundland and Labrador to whelp when Gulf ice conditions are poor.[180]

In East Greenland, the current size of the harp seal stock is about three times as high as it was in the 1970s (estimated at more than 600,000 animals, range 470,540–784,280)[181] . So, as for polar bears in the southern Davis Strait, there are lots of harp seal pups for East Greenland polar bears to eat in the spring and early summer. However, for unknown reasons, hooded seals in the same area (called the 'West Ice') appear to be declining (although those that live off Newfoundland and Labrador are doing well). There are only about half as many hooded seals in East Greenland now as there were in 1997, and many fewer than there were in the 1950s.[182] That is a huge *actual* decline, not a predicted one. The hooded seal in East Greenland was listed as 'vulnerable' to extinction by the IUCN in 2008 but, oddly, that fact hasn't been making headlines.

Chukchi Sea polar bears have been doing better in recent years, with an extended open-water season, than they did during the 1980s. This is because the ringed and bearded seals that are their primary prey do most of their feeding in ice-free summer waters.[183] More fat seals mean more fat seal pups the following spring for polar bears to eat. A recent study found that since 2007, with longer ice-free summers than during the 1980s, the summer feeding period for seals was extended and they became extra fat. Well-fed female ringed seals produced fat healthy pups the next spring, which means more food for polar bears when they need it the most. It seems likely this is the case in many other peripheral regions of the Arctic with wide continental shelves (such as the Laptev, Kara and Barents Seas) but not the Southern Beaufort. Oddly, in March 2013, less than six months after ringed and bearded seals were listed as 'threatened' with extinction, seal biologists were reporting to their peers that the results of their Chukchi Sea research contradicted their dire predictions: less summer sea ice was actually better for ringed and bearded seals, not worse.[184]

As the USA stands out as the only nation that insists Arctic seals will be harmed by future declines in summer sea ice, as noted above, it was all the more surprising that in September 2017, the US Fish and Wildlife Service announced it would not pursue a plan to list Pacific walrus as 'threatened' with extinction under the Endangered Species Act. Walrus experts concluded there was no evidence of ongoing harm or an imminent threat to walrus survival, a position which now concurs with the IUCN Red List assessment for this species.[185]

7 Health and survival

Body condition

There has been no increase in the percentage of starving or dying bears in recent years compared to the 1980s, despite the starving bear photos and videos that have gone viral on the internet. While such images have been used to make points about human-caused global warming and loss of Arctic sea ice, none of the photos circulated to date show bears unequivocally harmed due to reduced sea ice and lack of prey. One photo distributed in 2015 showed a Svalbard bear with a badly injured leg; a video that went viral of an emaciated Baffin Bay bear[186] almost certainly showed a bear suffering from cancer or another malady that left it unable to hunt, causing profound wasting and starvation. Baffin Bay bears normally spend the summer fasting on land, and in recent years the population has been doing well.

Female body condition of polar bears has been reported to be somewhat worse in a few areas (SB, SH, DS, BB), but not to below threshold levels necessary for reproduction.[187] A recent mark-recapture survey for WH did not report female body condition, which means this metric has not been updated since 2007.[188] However, in SH, Martin Obbard and colleagues determined that in the 2000s, females were on average about 31 kg lighter than they were in 1980s and males 45 kg lighter. However, the number of bears in the population did not decline over the same period, which suggests that the small decline in body condition reported was not significant to survival.[189] Previous research on WH bears captured between 1982 and 1990 stated that the critical weight for pregnant females was about 189 kg (below this weight, they lost the pregnancy).[190] Obbard and colleagues did not mention finding any mature females at or near this critical point, nor did any of the other reports that documented a decline in body condition. Considering that males can be over 500 kg and females over 300 kg by the time they come ashore in late summer,[191] it is doubtful that an average weight loss of 31–45 kg would have an appreciable effect on bear survival or reproduction in any subpopulation.

In contrast, a just-released paper reports that Southern Beaufort Sea females are having difficulty catching prey and notes a notable lack of seal pups in the diet in 2014–2016, resulting in uncharacteristic spring weight loss. This is circumstantial evidence that thick spring ice events that have impacted the region every ten years or so since the 1960s have continued, although authors Pagano and colleagues do not draw that conclusion.[192] The two most devastating periods of heavy spring ice occurred in 1974–1976 and 2004–2006, with similar but perhaps less profound events in the mid-1960s, 1984–1986, 1992–94.[193] If confirmed, a thick spring ice episode for the Southern Beaufort in 2015–2016 would be right on schedule, 10 years after the 2004–2006 events. Yet virtually all of the experts involved in assessing Southern Beaufort polar bears[194] seem to be blaming increased numbers of bears fasting in the spring of 2014–2016 on reduced summer sea ice,[195] as they did for starving and fasting bears, incidents of cannibalism, and population declines documented in 2004–2006.[196]

Effect of record low winter ice

As far as is known, the record low extent of sea ice in March 2017[197] had no impact on polar bear health or survival. Evidence for this position comes from ongoing research from the region around Svalbard, Norway by Norwegian Polar Institute biologists, who found no differences in body condition of adult male polar bears caught in the spring of 2017 (March–May) compared to those caught in previous years (back to 1993, results published online).[198]

Hybridization

Claims of recent widespread hybridization of polar bears with grizzlies, known for years from the Central Canadian Arctic,[199] were disproven in 2016 and 2017. DNA analysis showed that a heavily publicized claim of a polar bear × grizzly hybrid shot in Western Hudson Bay in 2016, supposedly a first for this region,[200] was actually a blonde-coloured grizzly.[201] Polar bear specialist Ian Stirling was quoted as saying,[202] 'I think it's 99 per cent sure that it's going to turn out to be a hybrid', but the DNA test proved him wrong.

A DNA study discovered that a number of hybrids reported in the region around Banks Island in the western Canadian Arctic were all related: they comprised eight offspring of a single polar bear female (now dead) that mated with two grizzly bear males.[203] Four of these offspring were second-generation backcross hybrids, the result of the mating of at least two fertile female hybrids (living on the ice like polar bears) with their own grizzly bear father; three of the known first-generation offspring are dead. So, rather than a widespread phenomenon, the spate of hybrid bears identified since 2006[204] derive from a single polar bear with 'atypical mating preferences'. The hybridization events happened because male grizzlies from the mainland invaded polar bear habitat by walking over the sea ice in spring, not because of reduced sea ice, as previously claimed by some.[205] Tundra grizzlies have been doing well because longer summers seem to be improving grizzly health and reproduction; they are expanding their range north and southeast[206] of areas they occupied decades ago and numbers appear to be increasing, although a comprehensive survey has not yet been done.[207]

Another recently published study that examined a very large sample of polar bear DNA and found no evidence of hybridization, and a bear evolution paper found hybridization was just as common between other bear species as between polar bears and grizzlies.[208] Altogether, this evidence should have put to rest the notion that global warming was leading to widespread hybridization of polar bears with grizzlies, but it did not. In fact, none of the above evidence has stopped some writers from perpetuating hybridization-caused-by-global-warming myths. The

January 2018 issue of *National Geographic* and the 10 February 2018 issue of *New Scientist* both repeat claims about grizzly/polar bear hybrids that disregard new evidence.[209]

Effect of contaminants

Contaminants have been shown to be present in sea ice but have not been shown to have done any harm. Most of the data are from Eastern Greenland, where there has never been a polar bear population count. Even if harm could be shown to have occurred, no impact on population size could be inferred. One researcher undertook a long and extensive review of all the toxicology research done on polar bears to that date[210] and noted: 'published polar bear data included in this review are correlative and descriptive and therefore do not directly demonstrate contaminant mediated cause and effect relationships.'

While it is true that some biological effects have been recorded for a number of substances (e.g. sizes of male (but not female) skulls, changes in gene function, reduced penis-bone density), it has not been demonstrated that any of the changes documented have negatively affected polar bear health or population size. For example, there is no evidence that any penis bones of polar bears in East Greenland have broken in recent years due to low bone density. There is only a suggestion that that this could, theoretically, happen at some time in the future if the trend in density continues.[211] In short, all of the so-called 'evidence' for negative effects of organic pollutants on East Greenland polar bears is currently circumstantial and inconclusive.[212]

Cannibalism

There have been no further reports of unusual numbers of deaths due to cannibalism since 2011.[213] In mid-August that year, a video showing a fat Baffin Bay male killing and eating a young cub accompanied by its mother was portrayed by the media as 'evidence of climate change'.[214] However, since such episodes are natural events that have always occurred, it was simply serendipitous that the incident was captured on camera; it was nothing to do with the plight of a starving bear.

Den collapses

There have also been no reports of deaths due to den collapses in winter or spring since 1989, although this was predicted to be a serious problem for polar bears in a warming world.[215]

Swimming bears

There have been no further reports of polar bear deaths due to drowning during the open water season since 2004, and no evidence has been presented to show that long-distance swims are detrimental to the health or survival of polar bears.[216] One researcher found that bears in Hudson Bay made few long-distance swims (>50 km) in 2007–2012, and 60% of those started on pack ice and ended on land during sea-ice breakup in July; more Beaufort Sea bears undertook swims than Hudson Bay bears, but 80% of BS swims took place before the September sea-ice minimum, and bears started and ended in the pack ice as they moved north with the retreating ice edge.[217] The media were impressed by the feat of a single long-distance swim made by a BS female and cub in 2008,[218] which was reported in 2011 and promoted again in 2017.[219] However, a comparison of the numbers show the female lost slightly less weight during her 63-day swim and subsequent walk over the ice (49 kg or 109 lbs) than a typical bear sitting on the shore of Western Hudson Bay in the summer (54 kg or 119 lbs or 0.85 kg per day).[220]

Denning on land

There has been no statistically significant change in proportion of Southern Beaufort females that make their dens on the sea ice (51%) versus on land or nearshore ice within 5 km of land (49%) between the mid-1980s and 2013, despite marked increases in the length of the ice-free season.[221] Karyn Rode and colleagues[222] examined factors that might have been responsible for the higher reproductive success of both Southern Beaufort and Chukchi Sea females that made their dens on land rather than sea ice, but considered only total time spent in dens, spring and autumn snowfall amounts, autumn ice conditions, and spring and autumn air temperatures: in other words, they looked at everything except sea ice thickness in spring or availability of newborn prey in spring, conditions which are known to have had a very strong negative effect on survival of bears in the Southern Beaufort from 2004–2006, almost certainly impacting nearshore or land-denning bears more than ice-denning bears.[223]

Ice-free period on land

In recent years, the SB has been virtually 100% covered by sea ice between June and November, and the majority of bears stay on the ice as it retreats north in the summer; only a small fraction (17.5%) stay on land.[224]

Litter sizes

Within the three Hudson Bay subpopulations, litter sizes estimated from recent autumn surveys varied only slightly (Table 2). No trends in autumn litter sizes over time were found for BB bears between 1997 and 2013, and mean 2011–2013 litter size (1.55) was similar to FB and SH: both populations are considered stable.[225] For Kane Basin, mean autumn litter size in 2012–2014 was similar to WH in 2016. Twin litters are common (Figure 9), but both don't always survive until autumn.[226]

8 Evidence of flexibility

Den locations

In the Barents Sea, where in some recent years the sea ice has not returned to the east coast of Svalbard in time for pregnant females to access traditional denning areas in autumn, it appears

Table 2: Litter sizes estimated from recent autumn surveys.

Subpopulation	Litter size	Year	Reference
WH	1.63	2016	Dyck et al. 2017
WH	1.43	2011	Dyck et al. 2017
WH	1.50	1985–1992	Derocher and Stirling 1995
WH	1.62	1980–1984	Derocher and Stirling 1995
WH	1.56	1966–1979	Derocher and Stirling 1995
SH	1.56	2011	Dyck et al. 2017
FB	1.54	2009/2010	Dyck et al. 2017
BB	1.55	2011–2013	SWG 2016: 301, 321
KB	1.60	2012–2014	SWG 2016: 552
KB	1.67	1992–1997	SWG 2016: 552

that the bears affected have been sufficiently flexible to use the much colder but still productive islands of the Franz Josef Land archipelago instead (Figure 8).[228]

In Baffin Bay, females in 2009–2015 entered land dens a bit later in the autumn than they did in 1991–1997 (about 5 October versus 5 September) and made dens at higher locations, but they emerged at similar times in both periods. However, no negative effects of these changes were noted.[229] This flexibility in choosing the best time and place to enter a den is evidence that

Figure 8: A fat pregnant female bear climbs a steep slope on Franz Josef Land in the Russian sector of the Barents Sea in summer.

The area has many alternative denning sites that can be used in low ice years by females that usually den on the east coast of Svalbard.[227] Vladimir Melnik photo, Shutterstock.

polar bear females have the biological flexibility necessary to survive changing environmental conditions.

Figure 9: A mother and two cubs-of-the-year on land in autumn.
Sergey Uryadnikov photo, Shutterstock.

Feeding locations

In 2013, fewer Baffin Bay females were traced moving south into Davis Strait in winter/spring to pursue harp and hooded seal than in 1997. Similarly, in summer, fewer bears visited Lancaster Sound, where there is often remnant sea ice to use as a hunting platform, than did so in the 1990s. More BB females in 2013 remained in the northern portion of their range during ice-covered seasons than they did previously.[230] These changes in distribution of female bears appear to relate to feeding behaviour. While the authors of the study attempted to correlated the changes with changes in sea ice coverage between the early 1990s and the 2009–2015 period, there appeared to be no attempt to consider potential changes in prey availability that may have taken place over that time.[231]

An older example of this kind of flexibility was the documented movement of bears and seals into the Chukchi Sea during the catastrophic 1974 and 1975 episodes of thick spring ice in the Eastern Beaufort.[232] In the 1960s, Christian Vibe also described seals and bears moving in response to decadal cycles of changes in sea ice cover along the Greenland coast.[233]

More recently, feeding on whale carcasses left over from Inuit subsistence hunting has been shown to benefit the small proportion of the Southern Beaufort subpopulation that spends the summer on shore rather than on the sea ice.[234] However, aside from whale carcasses, there is little evidence that terrestrial foods make a difference to the body condition or survival of bears that spend all or part of the summer onshore in the ice-free season.[235] While polar bears have been documented eating a variety of foods while onshore, from ground-nesting birds and their eggs to caribou, grasses, berries, and seaweed,[236] there is little evidence this makes any difference to body condition or survival over the short or long term.

Although the reasons for long distance moves are often not clear, they do happen. A four-year-old female who had not yet given birth travelled from the Canadian area of the Southern Beaufort Sea to Wrangel Island in the Chukchi Sea after being captured and tagged in late April 2009.[237] Previously, another bear, an adult female with two cubs of the year who was tagged in late May 1992, moved from off Prudhoe Bay in the Southern Beaufort Sea, crossed the Arctic Basin to within 2 degrees of the North Pole, and ended up in northern Greenland.[238]

Genetics

One recent, widely publicized genetics paper suggested that there is evidence that polar bears have already started moving from the periphery of the Arctic towards a sea ice 'refugium' in the Canadian Archipelago region in response to recent declines in summer sea ice. However, a follow-up analysis that did not get any media attention found 'methodological shortcomings' (including small and unbalanced sample sizes) and 'errors of interpretation' undermined the conclusions of the first study.[239] The second study did not find evidence of recent widespread movement towards the Canadian Archipelago but did confirm the existence of a genetically unique cluster of bears in Norwegian Bay previously identified by other researchers.[240] Norwegian Bay is located at the north end of the Canadian Archipelago and, while it is dominated by multiyear ice, it also has two large polynyas that have a few ringed seals and also support walrus and bearded seal populations.[241] Anecdotal accounts from local Inuit suggest that Norwegian Bay bears are 'different' from those in the surrounding area,[242] thus corroborating the two independent genetic studies. It is possible this subpopulation contains remnants of a previous population since gone into decline.[243]

One of the most recent genetic studies[244] emphasized that the polar bear, as a species, survived more than one previous warm period when there was virtually no summer sea ice.[245] Sea ice has varied both over the short term (i.e. decades-long climate oscillations) and the long term (glacial-to-interglacial cycles of thousands of years). Over the last 1.5 million years, for example, there have been periods of much less ice than today (including ice-free summers), but also periods with much more ice but no biological extinctions.[246]

Polar bear population numbers may have fluctuated up and down somewhat in conjunction with these sea-ice changes but the polar bear as a species survived, and so did all of the Arctic seal species they depended on for food, including Pacific walrus.[247] Their survival indicates that these Arctic marine mammals, in an evolutionary sense, have the necessary flexibility ('plasticisty' in the jargon) to survive in their highly variable habitat.[248] Although some have suggested that the low genetic diversity of polar bears makes them especially vulnerable to extinction,[249] there is little support for this notion in the scientific literature.[250]

9 Human–bear interactions

Attacks on humans

A major 2017 scientific summary of polar bear attacks on humans (1880–2014), authored by biologist James Wilder and colleagues,[251] concluded that such attacks are extremely rare. However, this may be because they essentially ignored attacks on Inuit and other indigenous people that live and hunt in the Arctic. By attempting to generate information that could be assessed with statistical methods, the authors ended up with data so skewed and incomplete that it does not provide a plausible assessment of the risk to humans of attacks by polar bears. Acknowledging that well-reported attacks on Europeans (or recorded by them) make up the bulk of the data used in the paper does not adequately address the weakness of the authors' conclusions.

This means that, except for well-reported incidents from the last few decades, virtually all attacks on the people most likely to encounter polar bears are not included in the study and the authors discount the almost perpetual danger from polar bear attack that Inuit and other indigenous people endured – and still endure in many areas – because those people in the past existed in 'relatively low numbers'. Oddly, no supplementary data is provided to show which records of attacks were included in the study, and no information is provided about how to access the database.

The paper focuses much attention on the potential for increases in polar bear attacks on humans due to sea-ice loss blamed on human-caused global warming,[252] but ignores totally the increased risk stemming from the larger proportion of adult males in populations that are now totally protected from hunting. Adult males frequently steal the kills of younger bears,[253] and in recovering (i.e. growing) populations, relatively more adult males are present to potentially dominate young bears and cause them to be nutritionally stressed and at risk of attacking humans. Ian Stirling warned in the early 1970s that a complete hunting ban, such as Norway had just imposed, might increase polar-bear–human conflicts.[254]

Problem bears in winter/spring

Winter is the leanest time of year for polar bears (Figure 7), since fat Arctic seal pups won't be available for another 2–3 months and meals for polar bears are hard to come by; this makes the bears especially dangerous when they come into contact with humans.[255] By early spring, bears are in hunting mode, as they pack on as much fat as possible to aid their survival over the summer months of fasting, and humans do well to avoid being the focus of these hunts.

Although winter incidents have been relatively rare over the last few decades, there are now many more polar bears than there were in the 1970s as well as more people living in many coastal Arctic communities. This means that problems with bears in winter are likely to increase. More bears out on the ice in winter (January–March) will almost certainly create more competition for the few seals that are available. As a result some bears might look for alternate sources of food. On land in winter, bears are attracted by caches of frozen meat, cemeteries, odours of cooking food, food fed to dogs and the dogs themselves, stored food, garbage and sewage, as well as man-made petroleum products and other industrial material (such as oils and lubricants, vinyl seats and plastic-coated cables), antifreeze and insulation.[256]

The incidents mentioned in the following subsections are anecdotal, and are not part of a comprehensive survey that would make them scientifically significant. However, until such a survey is undertaken, they are noted here to provide a perspective on the summer incidents and attacks on humans that often garner media attention.[257]

Svalbard 2017

In late January 2017, a polar bear female with two cubs (possibly two-year-olds) were reported near the community of Longyearbyen on the west coast of Svalbard, where there was no sea ice. The bears probably travelled overland from the east coast where spare sea ice existed. The bears were not reported to have been thin, starving, or in poor condition, and had thus far not caused any problems aside from frightening people. Helicopters and snowmobiles were used to chase them away from the community, but even so, the bears remained in the vicinity for several weeks.[258]

Greenland 2016

A polar bear killed and partially consumed a horse in the middle of February, in a community on the southwestern tip of Greenland,[259] weeks before newborn seals would be available in the pack ice. According to one report, this was the second time in two days that a polar bear was shot in Greenland because it got too close for comfort. More than one polar bear onshore in mid-winter in this region is rare or unheard of in recent memory (the odd sighting nearby, perhaps, but not bears prowling within communities). No age, sex or condition of the bear was provided in the report. Later in the year, in early April, a British cameraman reported fighting off a polar bear with a flare when it tried to smash into his cabin on the east coast of Greenland.[260]

East Hudson Bay/Hudson Strait 2017

A 'very fat' polar bear was reported outside the community of Inukjuak, Eastern Hudson Bay, on Saturday 25 February 2017.[261] This was a rare occurrence: according to the mayor, the community had not seen a bear onshore in nearly 30 years. The bear was a young, subadult female in excellent condition but it was shot for safety reasons. Its condition was surprising, as subadults are likely to be in poorer condition than adults at any time of year, due to their lack of hunting experience and competition with adult males.[262] Polar bears in Hudson Bay travel with the retreating ice to the western and southern shores so, with some exceptions, they usually only have access to the east coast during winter through spring.

Further up the east coast of Hudson Bay, a few weeks later, in early March, there was a late-night encounter with a thin and hungry polar bear in the northern Quebec community of Ivujivik on the edge of Hudson Bay. It was the fourth defence kill of 2017 (and the second that month), coming after a large number of bear sightings by residents that winter.[263] This bear was thin and obviously dangerous but was shot before anyone was hurt.

Labrador and Newfoundland 2016, 2017

In late January 2016, the tracks of a polar bear were spotted on Fogo Island close to the community of Tilting, but it apparently returned to the sea ice without any trouble.[264] However, later that year a large juvenile male polar bear was shot by an RCMP officer as it came ashore on the west coast of Fogo in early May.[265] Several sightings of bears were reported in and around Black Tickle on the southern Labrador coast in late January and early February, including a bear that peeked into the window while a man cooked dinner and another that took food put out for a resident's sled dogs.[266]

In 2017, there were more than a dozen sightings of polar bears onshore in Newfoundland and Labrador between January and May due to unusually cold conditions and heavy sea ice offshore.[267] While one or two bears have been reported in Newfoundland every year since about 2012,[268] there were at least half a dozen sightings in the spring of 2017,[269] perhaps the largest number of bears ashore since 1880.[270] The Davis Strait subpopulation was heavily impacted by commercial whalers in the late 1800s and early 1900s[271] and abundant prey in the form of harp and hooded seals[272] may only now be encouraging bears to wander to the southern limits of the region.

10 Discussion

It has come as a shock to many people that global polar bear numbers have been stable or slightly higher than they were in 2005, despite the fact that summer sea ice since 2007 has been at low levels not expected until mid-century. It seems hard to believe that the sea ice models used in 2005 to predict ice coverage over the 21st century could have been so flawed and that the polar bear survival models that predicted a 67% decline in abundance could have been so far off the mark.[273]

However, revised models for both now exist: new ice models predict nearly ice-free conditions in summer before mid-century[274] and a new polar bear survival model based on a new method of classifying sea ice suggests that by mid-century, a 30% decline or more in polar bear numbers should be expected – or at least that there is a 70% chance of such a decline.[275] In other words, experts now say we should expect an ice-free summer much earlier than previously predicted but can anticipate that 10,000 or so fewer polar bears will die because of it. That's a huge reversal since 2005 and a tacit admission that previous predictions were indeed unsound.

Recent studies have shown that abundant prey in the Chukchi Sea explains in part why global polar bear numbers did not decline as predicted, but adequate sea ice in spring and early summer since 2007 has likely been a contributing factor. This unexpected finding is almost certainly as true for unstudied 'Divergent' ice subpopulations as for the Chukchi Sea, with the exception of the Southern Beaufort. The Southern Beaufort is uniquely subject to periodic thick sea ice conditions in spring that cause temporary population fluctuations, which makes it an Arctic (and 'Divergent' ice) anomaly.

By far the greatest change in sea ice habitat since 1979 has been experienced by Barents Sea polar bears and the least by those in Southern Hudson Bay, the most southerly region inhabited by bears. Yet, surprisingly, polar bear numbers in the Barents Sea did not decline over the 2004–2015 period despite the loss of ice. Moreover, bear abundance in Southern Hudson Bay, with the least amount of ice loss, has been stable since the 1980s. The proclaimed correspondence between sea ice decline and population decline is certainly not evident in these data.

Claims of widespread hybridization of polar bears with grizzlies have been disproven by DNA studies published over the last few years, yet the media and some scientists perpetuate the unfounded claim that future sea ice loss threatens the integrity of the species through hybridization.

Starving bears have had vast amounts of attention in recent years. While some studies show bears are lighter in weight than they were in the 1980s, none showed an increase in the number of individuals starving to death or too thin to reproduce. The researchers did not come out and say so, but the early 2018 report about Southern Beaufort Sea bears having difficulty finding newborn seal prey in 2014–2016[276] strongly suggests that the thick ice events that have impacted the region every ten years or so since the 1960s have continued despite reduced summer sea ice. While the paper was not about starving bears, you would not have known it from the over-the-top media coverage, which also prompted a resurgence of the *National Geographic* Baffin Bay starving bear video that had gone viral in December 2017. The videographers responsible for the footage justified their actions by saying they wanted to show 'what a starving bear looks like', but that seemed inappropriate and manipulative to many viewers. Others were critical that no steps were taken to have the bear humanely put down.[277] Overall, the message intended by this video probably backfired, with a significant proportion of the public now understanding that starvation has always been the leading cause of death for polar bears of all ages.[278]

The media do not deserve all the blame for promoting polar bear disaster: it was the polar bear specialists who began the stories of looming catastrophe in the first place. Back in 2007, even as their report to support the US Fish and Wildlife proposal to list polar bears as 'threatened' with extinction under the Endangered Species Act (ESA) went to press in late August[279], polar bear biologist Steven Amstrup[280] knew that a precipitous sea-ice decline to 2050-like levels was imminent. The apprehension was palpable in Amstrup's report: 'the sea ice in 2007 already has declined below the level projected for mid century by the 4 most conservative models in our ensemble...'.[281] What the public heard was 'it's worse than we thought' and assumed polar bears were doomed. The media latched onto the message of imminent catastrophe conveyed by the biologists and ran with it.

Several years later, a 2010 academic paper discussed Amstrup's experience with publicizing the polar bear ESA listing decision:[282,†]

> He [Steven Amstrup] had been struggling with the dawning realization that the dire prognosis for polar bears that he and coworkers had issued had been perceived by the general public as a prediction of unavoidable doom for the species. He fired off a passionate e-mail

† e.g., Amstrup et al. 2008, Regehr et al. 2010.

to his colleagues, making a plea for hope. We quote some of this e-mail (with his permission) below:

> '...I was much chagrinned by the first flurry of reports in the media covering the release of our information. The take home message seemed to be that polar bears are going to disappear and there is nothing we can do about it.'

However, given that sea ice extent after 2007 remained below predicted levels, the message of disaster for polar bears became entrenched. The public, including Al Gore,[283] believed polar bear researchers in 2007 when they implied the bears were doomed. Amstrup and colleagues tried to mitigate the damage in a more hopeful and well-publicized 2010 paper in *Nature*,[284] which concluded things would be OK if immediate measures were taken to mitigate CO_2 emissions blamed for global warming. But as CO_2 continued to rise,[285] the public and the media seemed to accept that polar bears were headed for extinction.

Fast forward a decade to 2017 and polar bears did not get a mention in Al Gore's *Inconvenient Truth* sequel.[286] With global polar bear numbers inching gradually upwards, the message of doom for the species was falling flat. Summer sea-ice levels were below 5 million km^2 but polar bears were still going strong. It seemed the stuffing had been knocked out of the big white global warming icon. Based on current conditions, as long as spring and early summer ice remain abundant, it looks as though polar bears will continue to recover from the over-hunting of the early 20th century.

Notes

1. The author has published original research results in related fields (Crockford 1997a,b, 2016; Crockford et al. 1997, 2011; Koop et al. 2000; Olesiuk et al. 1990, Tollit et al. 2009; Wilson et al. 2011), research on evolutionary theory that includes polar bears (Crockford 2003a-b, 2004, 2006), research on evolutionary theory that includes geological and atmospheric processes (Crockford 2009), reviews, critiques, and synthesis reports on polar bears and walrus (Crockford 2008a, 2012a,b, 2014a,b, 2017a–e), reviews and synthesis reports on Arctic climate and seals (Crockford 2008b, 2015, Crockford and Frederick 2007, 2011), critiques/commentaries in related fields (Crockford 2002, Crockford and Kuzmin 2012; Rolland and Crockford 2005), and edited a volume of original research in a related field (Crockford 2000).

2. https://www.amazon.com/s/ref=dp_byline_sr_book_1?ie=UTF8&text=Susan+J+Crockford&search-alias=books&field-author=Susan+J+Crockford&sort=relevancerank and https://susancrockford.com.

3. Crockford, S. 2017. Testing the hypothesis that routine sea ice coverage of 3–5 million km^2 results in a greater than 30% decline in population size of polar bears (Ursus maritimus). PeerJ Preprints 2 March 2017. Doi: 10.7287/peerj.preprints.2737v3

4. Crockford 2014a–c, Crockford 2015a,b, Crockford 2017b–d.

5. Obbard et al. 2010.

6. Amstrup et al. 2007; US Fish & Wildlife 2008.

7. Wiig et al. 2015; Regehr et al. 2016.

8. Aars et al. 2017; Crockford 2017a; Crockford and Geist 2018; Dyck et al. 2017; SWG 2016.

9. Larsen 1972.

10. Crockford 2017a; Crockford and Geist 2018.

11. Amstrup et al. 2007; USGS 2017.

12. Hunter et al. 2007, 2010.

13. Crockford 2017a; Crockford and Geist 2018.

14. Obbard et al. 2015; Dyck et al. 2017.

15. https://news.nationalgeographic.com/2017/12/polar-bear-starving-arctic-sea-ice-melt-climate-change-spd/. [7 December 2017] ; http://nationalpost.com/news/canada/what-everybody-got-wrong-about-that-viral-video-of-a-starving-polar-bear [12 December 2017]; https://www.nationalgeographic.com/photography/through-the-lens/starving-polar-bear-photo-climate-change-canada-spd/ [16 January 2018]

16. Wiig et al. 2015 Regehr et al. 2016.

17. https://www.ec.gc.ca/nature/default.asp?lang=En&n=A997D1CC-1

18. http://www.businessinsider.com/us-canada-clash-over-polar-bear-status-2016-2.

19. Akçakaya et al. 2006; Crockford 2017; Aars et al. 2006.

20. Laidre et al. 2016.

21. Anonymous 1968.

22. http://pbsg.npolar.no/en/status/pb-global-estimate.html.

23. Laidre et al. 2016; Regehr et al. 2016; Wiig et al. 2015.

24. Stroeve et al. 2007.

25. Amstrup et al. 2007; Durner et al. 2007; USGS 2007.

26. Crockford 2017a; Crockford and Geist 2018.

27. Amstrup et al. 2007, 2008; Durner et al. 2007, 2009.

28. Atwood et al. 2016a; Regehr et al. 2016; Vongraven et al. 2012; Vongraven 2013; Wiig et al. 2015; but see Harris 2012.

29. Crockford 2017; DeMaster et al. 1980; Ramseier et al. 1975; Stirling et al. 1975a, 1975b, 1980, 1981, 1982, 1985, 1988; Stirling and Archibald 1977; Stirling and Lunn 1997; Stirling 1997, 2002.
30. Pagano et al. 2018; Rode et al. 2018; Stirling and Lunn 1997; Stirling 2002; Harwood et al. 2012, 2016; Bromaghin et al. 2015.
31. e.g. Smith 1987; Stirling and Lunn 1997.
32. e.g. Cherry et al. 2009; Harwood et al. 2012; Pilfold et al. 2012, 2014.
33. http://pbsg.npolar.no/en/status/populations/baffin-bay.html.
34. Taylor et al. 2005.
35. Obbard et al. 2010; York et al. 2016.
36. https://www.canada.ca/en/environment-climate-change/services/biodiversity/maps-sub-populations-polar-bears-protected.html#_fig04.
37. SWG 2016 Summary
38. Pers. Commun.
39. Stirling et al. 1980; Stirling and Kiliaan 1980.
40. http://pbsg.npolar.no/en/status/populations/davis-strait.htm
41. Peacock et al. 2013.
42. Rode et al. 2012.
43. DFO 2012, 2014; Kovacs 2015.
44. https://www.canada.ca/en/environment-climate-change/services/biodiversity/maps-sub-populations-polar-bears-protected.html#_fig04.
45. Obbard et al. 2010; Taylor et al. 2006.
46. Stapleton et al. 2012.
47. http://pbsg.npolar.no/en/status/populations/foxe-basin.html.
48. Derocher and Stirling 1995.
49. Wiig et al. 1995.
50. Regehr et al. 2007.
51. Lunn et al. 2016.
52. Stapleton et al. 2014.
53. http://pbsg.npolar.no/en/status/populations/western-hudson-bay.html.
54. Dyck et al. 2017.
55. Dyck et al. 2017, pp. 3, 37
56. Lunn et al. 1998; PBSG 2017.
57. Obbard et al. 2006, Obbard 2008.
58. Obbard et al. 2015.
59. Aars et al. 2006.
60. Aars et al. 2009.
61. Aars et al. 2009.
62. Aars et al. 2017, Table 3
63. Aars et al. 2017.
64. Crockford 2017.
65. http://polarview.met.no/.
66. http://pbsg.npolar.no/en/status/populations/barents-sea.html.
67. Matishov et al. 2014.
68. Wiig et al. 2005 supplement; Regehr et al. 2016.
69. http://pbsg.npolar.no/en/status/populations/kara-sea.html.

70. Derocher et al. 1998.
71. Aars et al. 2006.
72. SJC, personal archive
73. http://pbsg.npolar.no/en/status/populations/laptev-sea.html.
74. Wiig et al. 2005 supplement; Regehr et al. 2016.
75. Wiig et al. 2005 supplement; Regehr et al. 2016.
76. Aars et al. 2006: 34; Belikov 1993; Wiig et al. 1995: 24
77. Obbard et al. 2010.
78. PBSG 2013, Crockford personal archive
79. http://pbsg.npolar.no/en/status/populations/chukchi-sea.html.
80. Rode et al. 2013, 2018.
81. http://www.dailymail.co.uk/news/article-5110801/Polar-bears-scramble-mountain-feast-whale.html.
82. Ovsyanikov 2010; Ovsyanikov and Menyushina 2015.
83. http://www.dailymail.co.uk/news/article-5110801/Polar-bears-scramble-mountain-feast-whale.html.
84. Amstrup et al. 1986.
85. Amstup et al. 2005.
86. Amstrup et al. 2007; Regehr et al. 2006; Regehr et al. 2010, Hunter et al. 2007, 2010.
87. Cherry et al. 2009; Harwood et al. 2012; Stirling 2002; Stirling et al. 2008.
88. Bromaghin et al. 2013; Bromaghin et al. 2015.
89. http://pbsg.npolar.no/en/status/populations/southern-beaufort-sea.html.
90. Stirling et al. 2011.
91. http://pbsg.npolar.no/en/status/populations/northern-beaufort-sea.html.
92. e.g. Larson 1972.
93. Crockford, private archive
94. http://pbsg.npolar.no/en/status/pb-global-estimate.html.
95. http://pbsg.npolar.no/en/status/populations/east-greenland.html.
96. Wiig et al. 2015 supplement; Regehr et al. 2016.
97. Durner et al. 2009; Regehr et al. 2016; Vongraven et al. 2012.
98. http://pbsg.npolar.no/en/status/populations/barents-sea.html.
99. Durner and Amstrup 1993; van Meurs and Splettstoesser 2003.
100. Ovsyanikov 2010; Todd et al. 1992.
101. http://pbsg.npolar.no/en/status/populations/arctic-basin.html.
102. Arrigo et al. 2012; Gosselin et al. 1997; Lee and Whitledge 2005 vs. Pomeroy 1997.
103. Obbard et al. 2010; SWG 2016; York et al. 2016.
104. https://www.canada.ca/en/environment-climate-change/services/biodiversity/maps-sub-populations-polar-bears-protected.html#_fig02.
105. SWG 2016 summary: 21; Taylor et al. 2008.
106. SWG 2016 summary: 14
107. http://pbsg.npolar.no/en/status/populations/kane-basin.html.
108. Furnell and Schweinsburg 1984; Taylor et al. 2006.
109. http://pbsg.npolar.no/en/status/populations/mclintock-channel.html.
110. Taylor et al. 2002.
111. http://pbsg.npolar.no/en/status/populations/viscount-melville-sound.html.

112. Vongraven and Peacock 2011.
113. Taylor et al. 2009.
114. Peacock et al. 2013; Taylor and Lee 1995.
115. http://pbsg.npolar.no/en/status/populations/gulf-of-boothia.html.
116. Lunn et al. 2002; Stirling et al. 1984: 214; Taylor et al. 2008.
117. Honderich 2001.
118. http://pbsg.npolar.no/en/status/populations/lancaster-sound.html.
119. Taylor et al. 2008.
120. http://pbsg.npolar.no/en/status/populations/norwegian-bay.html.
121. Paetkau et al. 1999; Malenfant et al. 2016.
122. Amstrup et al. 2007; Folger 2018; Wang and Overland 2012.
123. https://www.canada.ca/en/environment-climate-change/services/biodiversity/maps-sub-populations-polar-bears-protected.html#_fig02.
124. See also Rode et al. 2013, 2014.
125. Bromaghin et al. 2015; Burns et al. 1975; Harwood et al. 2012; Crockford 2017; Stirling 2008.
126. Chambellant et al. 2012; Crockford 2014; Ferguson et al. 2005; Vibe 1965, 1967.
127. http://pbsg.npolar.no/en/status/status-table.html.
128. http://wwf.panda.org/_core/general.cfc?method=getOriginalImage&uImgID=%26%2AB%20%26%22N%5F6%0A
129. Aars et al. 2017.
130. SWG 2016.
131. Dyke et al. 2017; Lunn et al. 2016; Regehr et al. 2007; Stapleton et al. 2014.
132. Dyke et al. 2017.
133. Amstrup et al. 2007, 2008; Wiig et a. 2015; Regehr et al. 2016.
134. This problem is expected to remain as more subpopulations are surveyed or resurveyed with newer methods.
135. Wiig et al. 2015; Regehr et al. 2016.
136. See also Crockford 2017a
137. http://nsidc.org/arcticseaicenews/2017/04/another-record-but-a-somewhat-cooler-arctic-ocean/.
138. http://www.mosj.no/en/fauna/marine/polar-bear.html.
139. Perovich et al. 2017.
140. Walsh et al. 2017.
141. Stroeve et al. 2007.
142. http://nsidc.org/arcticseaicenews/.
143. Obbard et al. 2016.
144. Atwood et al. 2016a; Durner et al. 2009; Lang et al. 2017.
145. SWG 2017.
146. Pagano et al. 2018.
147. But see Rode et al. 2018a.
148. Burns et al. 1975; Bromaghin et al. 2015; Cherry et al. 2009; DeMaster 1980; Harwood et al. 2012, 2015; Pilfold et al. 2012, 2015, 2015; Rode et al. 2010, 2014; Smith 1987; Stirling 1997, 2002; Stirling et al. 1975a, 1975b, 1980, 1981, 1982, 1985, 1988, 1993, 2008; Stirling and Lunn 1997.
149. https://news.nationalgeographic.com/2018/02/polar-bears-starve-melting-sea-ice-global-warming-study-beaufort-sea-environment/.

150. Crockford 2017; Crockford and Geist 2018.
151. DeMaster 1980; Bromaghin et al. 2015; Ramseier et al. 1975; Stirling 2002, Stirling and Lunn 1997.
152. Crockford 2015b; Derocher et al. 2004; Vongraven et al. 2012; Laidre et al. 2008.
153. Stirling et al. 1982: 21.
154. Amstrup et al. 2007; Durner et al. 2007, 2009; Pilfold et al. 2017; Pongracz and Derocher 2007.
155. Cronin et al. 2014; Vibe 1965, 1967.
156. Crockford 2012b.
157. https://www.thegwpf.com/was-the-younger-dryas-cooling-event-caused-by-cosmic-impact-after-all/.
158. Moore et al. 2017; Young et al. 1997.
159. Cronin and Cronin 2015; Sha et al. 2016; Stein et al. 2017.
160. Regehr et al. 2016.
161. Stern and Laidre (2016) calculated similar metrics using a threshold of 50% ice cover, which is presumed to be ideal polar bear habitat, but the overall trends are the same.
162. Aars et al. 2017; Obbard et al. 2016.
163. Castro de la Guardia et al. 2017; Cherry et al. 2013; Obbard et al. 2015, 2016.
164. Castro de la Guardia 2017.
165. Castro de la Guardia et al. 2017; Lunn et al. 2016; Regehr et al. 2016; Stern and Laidre 2016.
166. Obbard et al. 2016.
167. Nowak 2003.
168. https://theconversation.com/in-the-path-of-the-polar-bears-what-its-like-to-be-an-arctic-scientist-47060.
169. Amstrup 2003; Derocher et al. 2002; Durner and Amstrup 1996; Hammill and Smith 1991; Knudsen 1978; Pilfold et al. 2015; Stirling 1975; Stirling and Øritsland 1995.
170. Hammill and Smith 1991.
171. Smith 1987; Smith and Hammill 1981; Smith et al. 1991.
172. Aars et al. 2017.
173. Burns 1970; Crockford et al. 2007; Crockford and Frederick 2011; Davis et al. 2008; Fedoseev 1975; Finley et al. 1983; Wiig et al. 1999.
174. Aars et al. 2017; Descamps et al. 2017; Fauchald et al. 2014.
175. Kelly et al. 2010; Cameron et al. 2010.
176. USFWS 2012a-b
177. Lowry 2016; Kovacs 2016b
178. DFO 2012, 2014; Kovacs 2015.
179. Stenson 2014; Stenson et al. 2015.
180. Sergeant 1991: 56
181. Stenson 2014.
182. http://www.iucnredlist.org/details/6204/0 and Kovacs 2016a
183. Crawford et al. 2015; Rode et al. 2018b
184. Crawford and Quakenbush 2013; USFWS 2012a-b
185. Lowry 2015; MacCracken et al. 2017; USFWS 2017.
186. https://news.nationalgeographic.com/2017/12/polar-bear-starving-arctic-sea-ice-melt-climate-change-spd/.
187. Obbard et al. 2016; Peacock et al. 2012; Rode et al. 2014, 2018b; SGW 2016.
188. Lunn et al 2016: 1315; Stirling and Derocher 2012: 2697.

189. Obbard et al. 2015, 2016.
190. Derocher et al. 1992.
191. Derocher et al. 1992.
192. Pagano et al. 2018; but see Harwood et al. 2015; Rode et al. 2018b
193. Bromaghin et al. 2015; DeMaster 1980; Melling 1996; Ramseier et al. 1975; Stirling 2002, Stirling and Lunn 1997.
194. Pagano et al. 2018; Rode et al. 2018b; Whiteman 2018.
195. https://news.nationalgeographic.com/2018/02/polar-bears-starve-melting-sea-ice-global-warming-study-beaufort-sea-environment/.
196. Amstrup et al. 2006; Bromaghin et al. 2015; Crockford 2017; Crockford and Geist 2018; Regehr et al. 2006, 2010; Rode et al. 2010, 2014; Stirling et al. 2008.
197. http://nsidc.org/arcticseaicenews/2017/04/another-record-but-a-somewhat-cooler-arctic-ocean/.
198. http://www.mosj.no/en/fauna/marine/polar-bear.html.
199. Doupé et al. 2007.
200. https://www.washingtonpost.com/news/animalia/wp/2016/05/23/love-in-the-time-of-climate-change-grizzlies-and-polar-bears-are-now-mating/ and http://www.cbc.ca/news/canada/north/grizzly-polar-bear-hybrid-arviat-nunavut-1.3586738.
201. http://www.nunatsiaqonline.ca/stories/article/65674exotic_bear_harvested_in_nunavut_was_a_blonde_grizzly/ and http://www.cbc.ca/news/canada/north/blonde-grizzly-hybrid-polar-1.3645547?cmp=rss.
202. https://www.thestar.com/news/canada/2016/05/23/did-nunavut-hunter-shoot-grizzly-polar-bear-hybrid.html.
203. Pongracz et al. 2017.
204. http://www.gi.alaska.edu/AlaskaScienceForum/article/hybrid-grizzly-polar-bear-curiosity.
205. Doupé et al. 2007; Kelly et al. 2010; Pongacz et al. 207; Post et al. 2013.
206. http://www.sciencedaily.com/releases/2010/02/100223121439.htm.
207. COSEWIC 2012; McLoughlin et al. 2003; Nielson et al. 2013.
208. Kumar et al. 2017; Peacock et al. 2015.
209. Folger 2018: 109; Pearce, 2018.
210. Sonne 2010: 468
211. Sonne et al. 2015.
212. Dietz et al. 2013a, 2013b; Gabrielson et al. 2015; McKinney et al. 2013; Sonne 2010; Sonne et al. 2015.
213. Amstrup et al. 2006; Derocher and Wiig 1999; Lunn and Stenhouse 1985; Stirling and Ross 2011; Taylor et al. 1985.
214. http://news.nationalgeographic.com/2016/02/160223-polar-bears-arctic-cannibals-animals-science/.
215. Clarkson and Irish 1991; Stirling and Derocher 2012, Derocher et al. 2004.
216. Monnett and Gleason 2006; Pilfold et al. 2017.
217. Pilfold et al. 2017.
218. https://www.washingtonpost.com/news/speaking-of-science/wp/2016/04/21/melting-arctic-ice-is-forcing-polar-bears-to-swim-for-more-than-a-week-without-rest/?utm_term=.392a416ffcdd.
219. Durner et al. 2011; reexamined by Griffen 2018.
220. Derocher and Stirling 1995.

221. Olson et al. 2017; Rode et al. 2018; Stern and Laidre 2016.
222. Rode et al. 2018a
223. Bromaghin et al. 2015; Crockford 2017; Harwood et al. 2012, 2015, Nyugen et al. 2017; Stirling et al. 2008, Amstrup et al. 2006.
224. Atwood et al. 2016; Olson 2017: 213
225. Obbard et al. 2015; Stapleton et al. 2016; SWG 2016: 316, 325
226. Derocher and Stirling 1995.
227. http://pbsg.npolar.no/en/status/populations/barents-sea.html.
228. Aars 2015; Aars et al. 2017; Barr 1995; Chernova et al. 2014; Descamps et al. 2017; Fauchald et al. 2014; Larsen 1972; PBSG 2017.
229. Escajeda et al. 2018.
230. Laidre et al. 2018.
231. DFO 2012; Laidre and Stern 2016; Sergeant 1976.
232. Burns et al. 1975; Ramseier et al. 1975; Smith and Stirling 1978; Stirling et al. 1975a, b
233. Vibe 1965, 1967.
234. Atwood et al. 2016b; Herreman and Peacock 2013; McKinney et al. 2017a; Miller et al 2006; Rode et al. 2015; Rogers et al. 2015.
235. Rode et al. 2015.
236. Iles et al. 2013; Iverson et al. 2014; Gormezano and Rockwell 2013a, 2013b; Tartu et al. 2016.
237. Johnson et al. 2017.
238. Durner and Amstrup 1995.
239. Peacock et al. 2015; Malefant et al. 2016.
240. Paetkau et al. 1999.
241. Stirling et al. 1981; Stirling 1997; Taylor et al. 2001.
242. Taylor et al. 2001: 704
243. Lindqvist et al. 2010; Miller et al. 2012.
244. Cronin et al. 2014.
245. https://www.adn.com/our-alaska/article/dna-study-pushes-back-date-divergence-polar-brown-bears/2014/04/01/.
246. Cronin and Cronin 2015.
247. Crockford 2012, 2014.
248. Abbitt and Scott 2001.
249. https://www.adn.com/our-alaska/article/dna-study-pushes-back-date-divergence-polar-brown-bears/2014/04/01/.
250. Aguillar et al. 2004; Hoelzel et al. 1993; Frasier et al. 2013; Kaeffer et al. 2007; Lehman 1998:R723–724; Lindqvist et al. 2010; Weber et al. 2004.
251. Wilder et al. 2017.
252. e.g. https://www.adn.com/arctic/2017/07/11/as-sea-ice-gets-scarcer-polar-bear-attacks-on-people-become-more-frequent/.
253. Stirling 1974.
254. Anonymous 1976: 11
255. Fleck and Herrero 1988; Gjertz et al. 1993; Gjertz and Persen 1987; International Human-Bear Conflicts Workshop 2009; Stenhouse et al. 1988.
256. Truett 1993.

257. http://www.dailymail.co.uk/news/article-2684735/I-tried-gouge-polar-bears-eyes-Trip-leader-Norway-expedition-Eton-schoolboy-tragically-killed-tells-horror-night-time-attack.html.
258. http://icepeople.net/2017/01/17/alert-polar-bear-tracks-spotted-at-unis-early-tuesday-morning-storm-making-monitoring-animals-difficult/.
259. http://arcticjournal.com/culture/2162/scare-tactics.
260. http://www.dailymail.co.uk/news/article-3584907/British-adventurer-reveals-fought-polar-bear-flare-animal-tried-smash-cabin-Greenland.html.
261. http://www.cbc.ca/news/canada/north/polar-bear-inukjuak-quebec-cabin-1.4004735.
262. Amstrup 2003.
263. http://www.nunatsiaqonline.ca/stories/article/65674nunavik_community_receives_some_unwelcome_guests/.
264. http://www.cbc.ca/news/canada/newfoundland-labrador/polar-bear-fogo-island-tilting-1.3421596.
265. http://www.cbc.ca/news/canada/prince-edward-island/polar-bear-rcmp-p-e-i-shaun-coady-1.3565340.
266. http://www.cbc.ca/news/canada/newfoundland-labrador/polar-bear-encounters-labrador-1.3432312.
267. http://www.cbc.ca/news/canada/newfoundland-labrador/ice-north-east-coast-newfoundland-canadian-coast-guard-1.4131418.
268. http://www.thestar.com/news/canada/2012/04/06/polar_bear_encounters_on_the_rise_in_newfoundland_and_labrador.html.
269. http://www.cbc.ca/news/canada/newfoundland-labrador/pack-ice-polar-bears-1.4065930.
270. Packard 1886; Stirling and Kiliaan 1980; Smith et al. 1975.
271. Honderich 1991.
272. Kovacs 2015; DFO 2012, 2014; Stenson 2014.
273. Crockford 2017; Crockford and Geist 2018.
274. Folger 2018; Overland and Wang 2013; Wang and Overland 2012; see also Barnhart et al. 2015.
275. Regehr et al. 2016; Stern and Laidre 2016.
276. Pagano et al. 2016.
277. https://www.nationalgeographic.com/photography/through-the-lens/starving-polar-bear-photo-climate-change-canada-spd/ [16 January 2018].
278. Amstrup 2003.
279. Amstrup et al. 2007.
280. Now at Polar Bears International
281. Amstrup et al. 2007: 35
282. Swaisgood and Sheppard (2010: 627)
283. http://www.abc.net.au/mediawatch/transcripts/s1887890.htm.
284. Amstrup et al. 2010.
285. https://www.theguardian.com/environment/2013/nov/06/co2-levels-hit-record-high.
286. http://www.smh.com.au/entertainment/movies/an-inconvenient-sequel--truth-to-power-review-gore-shows-value-of-persistence-20170807-gxquoi.html.

Bibliography

Aars, J. 2015. *Research on polar bears at Norwegian Polar Institute*. Online seminar ('webinar"), January 14.

Aars, J., Lunn, N.J. and Derocher, A.E. 2006. *Polar Bears: Proceedings of the 14th Working Meeting of the IUCN/SSC Polar Bear Specialist Group, Seattle, Washington, 20–24 June 2005.* Occasional Paper of the IUCN Species Survival Commission 32. IUCN, Gland, Switzerland.

Aars, J., Marques, T.A., Buckland, S.T., Andersen, M., Belikov, S., Boltunov, A., et al. 2009. Estimating the Barents Sea polar bear subpopulation. *Marine Mammal Science* 25: 35–52.

Aars, J., Marques,T.A, Lone, K., Anderson, M., Wiig, Ø., Fløystad, I.M.B., Hagen, S.B. and Buckland, S.T. 2017. The number and distribution of polar bears in the western Barents Sea, *Polar Research* 36: 1, 1374125.

Abbitt, R.J.F., and Scott, J.M. 2001. Examining differences between recovered and declining endangered species. *Conservation Biology* 15: 1274–1284.

ACIA 2005. *Arctic Climate Impact Assessment: Scientific Report.* Cambridge University Press.

Aguilar, A., Roemer, G., Debenham, S., Binns, M., Garcelon, D., and Wayne, R.K. 2004. High MHC diversity maintained by balancing selection in an otherwise genetically monomorphic mammal. *Proceedings of the National Academy of Sciences USA* 101: 3490–3494.

Akçakaya, H.R., Butchart, S.H.M., Mace, G.M., Stuart, S.N. and Hilton-Taylor, C. 2006. Use and misuse of the IUCN Red List criteria in projecting climate change impacts on biodiversity. *Global Change Biology* 12: 2037–2043.

Amstrup, S.C. 2003. Polar bear (*Ursus maritimus*). In *Wild Mammals of North America*, G.A. Feldhamer, B.C. Thompson and J.A. Chapman (eds), pp. 587–610. Johns Hopkins University Press, Baltimore.

Amstrup, S.C. and Durner, G.M. 1995. Survival rates of radio-collared female polar bears and their dependent young. *Canadian Journal of Zoology* 73: 1312–1322.

Amstrup, S.C., Durner, G.M., Stirling, I. and McDonald, T.L. 2005. Allocating harvests among polar bear stocks in the Beaufort Sea. *Arctic* 58: 247–259.

Amstrup, S.C., Marcot, B.G. and Douglas, D.C. 2007. Forecasting the rangewide status of polar bears at selected times in the 21st century. Administrative Report, US Geological Survey. Reston, Virginia.

Amstrup, S.C., DeWeaver, E.T., Douglas, D.C., Marcot, B.G., Durner, G.M., Bitz, C.M., and Bailey, D.A. 2010. Greenhouse gas mitigation can reduce sea-ice loss and increase polar bear persistence. *Nature* 468: 955–958.

Amstrup, S.C., Marcot, B.G., and Douglas, D.C. 2008. A Bayesian network modeling approach to forecasting the 21st century worldwide status of polar bears. pp. 213–268 in *Arctic Sea Ice Decline: Observations, Projections, Mechanisms, and Implications,* E.T. DeWeaver, C.M. Bitz, and L.B. Tremblay (eds.). Geophysical Monograph 180. American Geophysical Union, Washington, D.C.

Amstrup, S.C., Stirling, I. and Lentfer, J.W. 1986. Past and present status of polar bears in Alaska. *Wildlife Society Bulletin* 14: 241–254.

Amstrup, S.C., Stirling, I., Smith, T.S., Perham, C. and Thiemann, B.W. 2006. Recent observations of intraspecific predation and cannibalism among polar bears in the Southern Beaufort Sea. *Polar Biology* 29: 997–1002.

Amstrup, S.C. 2011. Polar bears and climate change: certainties, uncertainties, and hope in a warming world. Pgs. 11–20 in R.T. Watson, T.J. Cade, M. Fuller, G. Hunt, and E. Potapov (eds.), *Gyrfalcons and Ptarmigan in a Changing World, Volume 1.* The Peregrine Fund, Boise, Idaho.

Andersen, M. and Aars, J. 2016. Barents Sea polar bears (*Ursus maritimus*): population biology and anthropegenic threats. *Polar Research* 35: 26029.

Anonymous. 1968. Proceedings of the First International Scientific Meeting on the Polar Bear, 6–10 September 1965 at Fairbanks, Alaska. U.S. Dept. of the Interior (Resource Publication 16) and the University of Alaska (International Conference Proceedings Series, No. 1). http://pbsg.npolar.no/en/meetings/

Anonymous. 1976. Polar Bears: Proceedings of the 5th meeting of the Polar Bear Specialist Group IUCN/SSC, 3–5 December, 1974, Le Manoir, St. Prex, Switzerland. Gland, Switzerland IUCN. http://pbsg.npolar.no/en/meetings/

Anonymous. 1985. Polar Bears: Proceedings of the 8th meeting of the Polar Bear Specialist Group IUCN/SSC, 15–19 January, 1981, Oslo, Norway. Gland, Switzerland and Cambridge UK, IUCN. http://pbsg.npolar.no/en/meetings/

Arrigo, K.R., Perovich, D.K., Pickart, R.S., Brown, Z.W., van Dijken, G.L., Lowry, K.E., Mills, M.M., Palmer, M.A., Balch, W.M., Bahr, F., Bates, N.R., Benitez-Nelson, C., Bowler, B., Brownlee, E., Ehn, J.K., Frey, K.E., Garley, R., Laney, S.R., Lubelczyk, L., Mathis, J., Matsuoka, A., Mitchell, B.G., Moore, G.W.K., Ortega-Retuerta, E., Pal, S., Polashenski, C.M., Reynolds, R.A., Schieber, B., Sosik, H.M., Stephens, M., and Swift, J.H. 2012. Massive phytoplankton blooms under Arctic sea ice. *Science* 336: 1408.

Atwood, T.C., Marcot, B.G., Douglas, D.C., Amstrup, S.C., Rode, K.D., Durner, G.M. et al. 2016a. Forecasting the relative influence of environmental and anthropogenic stressors on polar bears. *Ecosphere* 7(6): e01370.

Atwood, T.C., Peacock, E., McKinney, M.A., Lillie, K., Wilson, R., Douglas, D.C., and others. 2016b. Rapid Environmental Change Drives Increased Land Use by an Arctic Marine Predator. PLoS ONE 11(6): e0155932.

Barnhart, K.R., Miller, C.R., Overeem, I. and Kay, J.E. 2016. Mapping the future expansion of Arctic open water. *Nature Climate Change* 6: 280–285.

Barr, S. 1995. *Franz Josef Land.* Oslo: Norwegian Polar Institute.

Boveng, P.L., Bengtson, J.L., Buckley, T.W., Cameron, M.F., Dahle, S.P., Megrey, B.A., Overland, J.E., and Williamson, N.J. 2008. Status review of the ribbon seal (*Histriophoca fasciata*). NOAA Technical Memorandum NMFS-AFSC-191.

Boveng, P.L., Bengtson, J.L., Buckley, T.W., Cameron, M.F., Dahle, S.P., Kelly, B.P. Megrey, B.A., Overland, J.E., and Williamson, N.J. 2009. Status review of the spotted seal (Phoca largha). NOAA Technical Memorandum NMFS-AFSC-200.

Bromaghin, J.F., McDonald, T.L. and Amstrup, S.C. 2013. Plausible combinations: An improved method to evaluate the covariate structure of Cormack-Jolly-Seber mark-recapture models. *Open Journal of Ecology* 3: 11–22

Bromaghin, J.F., McDonald, T.L., Stirling, I., Derocher, A.E., Richardson, E.S., Regehr, E.V., Douglas, D.C., Durner, G.M., Atwood, T. and Amstrup, S.C. 2015. Polar bear population dynamics in the Southern Beaufort Sea during a period of sea ice decline. *Ecological Applications* 25(3): 634–651.

Burns, J.J. 1970. Remarks on the distribution and natural history of pagophilic pinnipeds in the Bering and Chukchi Seas. *Journal of Mammalogy* 51(3): 445–454.

Burns, J.J., Fay, F.H., and Shapiro, L.H. 1975. The relationships of marine mammal distributions, densities, and activities to sea ice conditions (Quarterly report for quarter ending September 30, 1975, projects #248 and 249). *In* Environmental Assessment of the Alaskan Continental Shelf, Principal Investigators' Reports, July-September 1975, Volume 1. NOAA Environmental Research Laboratories, Boulder, Colorado. pp. 77–78.

Cameron, M.F., Bengtson, J.L., Boveng, J.K., Jansen, J.K., Kelly, B.P., Dahle, S.P., Logerwell, E.A., Overland, J.E., Sabine, C.L., Waring, G.T. and Wilder, J.M. 2010. Status review of the bearded seal (*Erignatha barbatus*). NOAA Technical Memorandum NMFS-AFSC-211.

Castro de la Guardia, L., Myers, P.G., Derocher, A.E., Lunn, N.J., van Scheltinga, A.D. 2017. Sea ice cycle in western Hudson Bay, Canada, from a polar bear perspective. *Marine Ecology Progress Series* 564: 225–233.

Chambellant, M., Stirling, I., Gough, W.A. and Ferguson, S.H. 2012. Temporal variations in Hudson Bay ringed seal (*Phoca hispida*) life-history parameters in relation to environment. *Journal of Mammalogy* 93: 267–281.

Chernova NV, Friedlander AM, Turchik A, Sala E. 2014. Franz Josef Land: extreme northern outpost for Arctic fishes. *PeerJ* 2:e692 https://peerj.com/articles/692/.

Cherry, S.G., Derocher, A.E. and Lunn, N.J. 2016. Habitat-mediated timing of migration in polar bears: an individual perspective. *Ecology and Evolution* 6(14): 5032–5042.

Cherry, S.G., Derocher, A.E., Stirling, I. and Richardson, E.S. 2009. Fasting physiology of polar bears in relation to environmental change and breeding behavior in the Beaufort Sea. *Polar Biology* 32: 383–391.

Cherry, S.G., Derocher, A.E., Thiemann, G.W., Lunn, N.J. 2013. Migration phenology and seasonal fidelity of an Arctic marine predator in relation to sea ice dynamics. *Journal of Animal Ecology* 82: 912–921.

Clarkson, P.L. and Irish, D. 1991. Den collapse kills female polar bear and two newborn cubs. *Arctic* 44: 83–84.

COSEWIC. 2012. COSEWIC assessment and status report on the Grizzly Bear *Ursus arctos* in Canada. *Committee on the Status of Endangered Wildlife in Canada*. Ottawa.

Crawford, J. and Quakenbush, L. 2013. *Ringed seals and climate change: early predictions versus recent observations in Alaska*. Oral presentation by Justin Crawfort, 28th Lowell Wakefield Fisheries Symposium, March 26–29. Anchorage, AK. http://seagrant.uaf.edu/conferences/2013/wakefield-arctic-ecosystems/program.php.

Crawford, J.A., Quakenbush, L.T. and Citta, J.J. 2015. A comparison of ringed and bearded seal diet, condition and productivity between historical (1975–1984) and recent (2003–2012) periods in the Alaskan Bering and Chukchi seas. *Progress in Oceanography* 136: 133–150.

Crockford, S.J. 1997a. *Osteometry of Makah and Coast Salish Dogs*, Archaeology Press #22, Simon Fraser University, Burnaby, B.C.

Crockford, S.J. 1997b. Archaeological evidence of large northern bluefin tuna, *Thunnus thynnus*, in coastal waters of British Columbia and northern Washington. *Fishery Bulletin* 95: 11–24.

Crockford, S.J. (ed.) 2000. *Dogs Through Time: An Archaeological Perspective.* Proceedings of the 1st ICAZ Symposium on the History of the Domestic Dog. Oxford, British Archaeological Reports (B.A.R.), Archaeopress S889.

Crockford, S.J. 2002. Animal domestication and heterochronic speciation: the role of thyroid hormone. pp. 122–153. *In:* N. Minugh-Purvis and K. McNamara (eds.) *Human Evolution Through Developmental Change*. Baltimore, Johns Hopkins University Press.

Crockford, S.J. 2003a. Commentary: Thyroid hormones in Neanderthal evolution: A natural or a pathological role? *Geographical Review* 92(1): 73–88.

Crockford, S.J. 2003b. Thyroid hormone phenotypes and hominid evolution: a new paradigm implicates pulsatile thyroid hormone secretion in speciation and adaptation changes. *International Journal of Comparative Biochemistry and Physiology Part A* 135(1): 105–129.

Crockford, S.J. 2004. Animal Domestication and Vertebrate Speciation: A Paradigm for the Origin of Species. Ph.D. dissertation. University of Victoria, Canada.

Crockford, S.J. 2006. *Rhythms of Life: Thyroid Hormone and the Origin of Species.* Trafford, Victoria.

Crockford, S.J. 2008a. *Some Things We Know – and Don't Know – About Polar Bears.* Science and Public Policy Institute (SPPI).

Crockford, S.J. 2008b. Be careful what you ask for: archaeozoological evidence of mid-Holocene climate change in the Bering Sea and implications for the origins of Arctic Thule. In *Islands of Inquiry: Colonisation, Seafaring and the Archaeology of Maritime Landscapes,* 113–131. G. Clark, F. Leach and S. O'Connor (eds.). Terra Australis 29 ANU EPress, Canberra.

Crockford, S.J. 2009. Evolutionary roots of iodine and thyroid hormones in cell-cell signaling. *Integrative and Comparative Biology* 49: 155–166.

Crockford, S.J. 2012a. *A History of Polar Bears, Ringed Seals, and other Arctic and North Pacific Marine Mammals over the Last 200,000 Years.* A report Prepared for The State of Alaska Department of Commerce, Community and Economic Development and The University of Alaska Fairbanks. Pacific Identifications Inc., Victoria, B.C.

Crockford, S.J. 2012b. *Annotated Map of Ancient Polar Bear Remains of the World.* Spotted Cow Presentations Inc., Victoria, British Columbia. Available at http://polarbearscience/references/.

Crockford, S.J. 2014a. On the Beach: Walrus Haulouts are Nothing New. *Global Warming Policy Foundation Briefing Paper 11*. http://www.thegwpf.org/susan-crockford-on-the-beach-2/.

Crockford, S.J. 2014b. The Walrus Fuss: Walrus Haulouts are Nothing New http://www.thegwpf.org/gwpftv/?tubepress_item=cwaAwsS2OOY&tubepress_page=2

Crockford, S.J. 2014c. Healthy Polar Bears, Less than Healthy Science. Global Warming Policy Foundation Note 10. London.

Crockford, S.J. 2015a. Twenty Good Reasons not to Worry about Polar Bears. Global Warming Policy Foundation Briefing Paper 14. London.

Crockford, S.J. 2015b. The Arctic Fallacy: Sea Ice Stability and the Polar Bear. Global Warming Policy Foundation Briefing Paper 16. London. Available at http://www.thegwpf.org/susan-crockford-the-arctic-fallacy-2/.

Crockford, S. 2016. Prehistoric mountain goat Oreamnos americanus mother lode near Prince Rupert, British Columbia and implications for the manufacture of high-status ceremonial goods. Journal of Island and Coastal Archaeology. ?

Crockford, S. 2017a. Testing the hypothesis that routine sea ice coverage of 3–5 million km^2 results in a greater than 30% decline in population size of polar bears (Ursus maritimus). *PeerJ Preprints* 2 March 2017.

Crockford, S. 2017b. *Polar Bear Scare Unmasked: The Saga of a Toppled Global Warming Icon.* Youtube video, February, Global Warming Policy Foundation http://www.thegwpf.org/gwpf tv/?tubepress_item=z6bcCTFnGZ0.

Crockford, S. 2017c. *The Death of a Climate Icon.* Youtube video, August, Global Warming Policy Foundation http://www.thegwpf.org/gwpftv/?tubepress_item=XCzwFall8OQ.

Crockford S.J. 2017d. Twenty good reasons not to worry about polar bears: an update. *Global Warming Policy Foundation Briefing Paper 28*.

Crockford, S.J. 2017e. *Polar Bears: Outstanding Survivors of Climate Change*. Amazon CreateSpace https://www.amazon.com/dp/1541139712/.

Crockford, S. and Frederick, G. 2007. Sea Ice Expansion in the Bering Sea during the Neoglacial: Evidence from Archaeozoology. *The Holocene* 17: 699– 706.

Crockford, S.J. and Frederick, G. 2011. Neoglacial sea ice and life history flexibility in ringed and fur seals. pp. 65–91. In, T. Braje and R. Torrey, eds. *Human and Marine Ecosystems: Archaeology and Historical Ecology of Northeastern Pacific Seals, Sea Lions, and Sea Otters.* U. California Press, LA.

Crockford, S.J., Frederick, G. and R. Wigen. 1997. A humerus story: albatross element distribution from two Northwest Coast sites, North America. *International Journal of Osteoarchaeology* 7: 378.1–5.

Crockford, S.J. and Kuzmin, Y.V. 2012. Comments on Germonpré et al., *Journal of Archaeological Science* 36, 2009 'Fossil dogs and wolves from Palaeolithic sites in Belgium, the Ukraine and Russia: osteometry, ancient DNA and stable isotopes', and Germonpré, Lázkičková-Galetová, and Sablin, *Journal of Archaeological Science* 39, 2012 'Palaeolithic dog skulls at the Gravettian Předmostí site, the Czech Republic'. *Journal of Archaeological Science* 39: 2797–2801.

Crockford, S.J., Moss, M.L., and J.F. Baichtal. 2011. Pre-contact dogs from the Prince of Wales archipelago, Alaska. *Alaska Journal of Anthropology* 9(1): 49–64.

Cronin, T.M. and Cronin, M.A. 2015. Biological response to climate change in the Arctic Ocean: the view from the past. *Arktos* 1: 1–18.

Cronin, M.A., Rincon, G., Meredith, R.W., MacNeil, M.D., Islas-Trejo, A., Cánovas, A. and Medrano, J.F. 2014. Molecular phylogeny and SNP variation of polar bears (*Ursus maritimus*), brown bears (*U. arctos*), and black bears (*U. americanus*) derived from genome sequences. *Journal of Heredity* 105: 312–323.

Davis, C.S., Stirling, I., Strobeck, C. and Coltman, D.W. 2008. Population structure of ice-breeding seals. *Molecular Ecology* 17: 3078– 3094.

DeMaster, D.P., Kingsley, M.C.S. and Stirling, I. 1980. A multiple mark and recapture estimate applied to polar bears. *Canadian Journal of Zoology* 58: 633–638.

Department of Fisheries and Oceans Canada (DFO) 2012. *Current status of Northwest Atlantic harp seals (Pagophilus groenlandicus).* Science Advisory Report 2011/070.

Department of Fisheries and Oceans Canada (DFO) 2014. *Status of Northwest Atlantic harp seals, (Pagophilus groenlandicus)*. DFO Can. Sci. Advis. Sec. Sci. Advis. Rep. 2014/011.

Derocher, A.E. 1999. Latitudinal variation in litter size of polar bears: ecology or methodology? *Polar Biology* 22: 350–356.

Derocher 2005. Population ecology of polar bears at Svalbard, Norway. *Population Ecology* 47: 267–275.

Derocher, A.E., Garner, G.W., Lunn, N.J., and Wiig, Ø. (eds.) 1998. Polar Bears: Proceedings of the 12th meeting of the Polar Bear Specialist Group IUCN/SSC, 3–7 February, 1997, Oslo, Norway. Gland, Switzerland and Cambridge UK, IUCN.

Derocher, A.E. and Stirling, I. 1995. Temporal variation in reproduction and body mass of polar bears in western Hudson Bay. *Canadian Journal of Zoology* 73: 1657–1665.

Derocher, A.E. and Wiig, Ø. 1999. Infanticide and cannibalism of juvenile polar bears (*Ursus maritimus*) in Svalbard. *Arctic* 52: 307–310.

Derocher, A.E., Wiig, Ø., and Andersen, M. 2002. Diet composition of polar bears in Svalbard and the western Barents Sea. *Polar Biology* 25(6): 448–452

Derocher, A.E., Lunn, N.J. and Stirling, I. 2004. Polar bears in a warming climate. *Integrative and Comparative Biology* 44: 163–176.

Descamps, S., Aars, J., Fuglei, E., Kovacs, K.M., Lydersen, C., Pavlova, O., Pedersen, Å.Ø., Ravolainen, V. and Strøm, H. 2017. Climate change impacts on wildlife in a High Arctic archipelago – Svalbard, Norway. *Global Change Biology* 23: 490–502.

Dietz, R., Rigét, F.F., Sonne, C., Born, E.W., Bechshøft, T., McKinney, M.A. and Letcher, R.J. 2013a. Three decades (1983–2010) of contaminant trends in East Greenland polar bears (*Ursus maritimus*). Part 1: Legacy organochlorine contaminants. *Environment International* 59: 485–493.

Dietz, R., Rigét, F.F., Sonne, C., Born, E.W., Bechshøft, T., McKinney, M.A., Drimmie, R.J., Muir, D.C.G., and Letcher, R.J. 2013b. Three decades (1984–2010) of contaminant trends in East Greenland polar bears (*Ursus maritimus*): Part 2 Brominated flame retardants. *Environment International* 59: 494–500

Doupé, J.P., England, J.H., Furze, M. and Paetkau, D. 2007. Most northerly observation of a grizzly bear (*Ursus arctos*) in Canada: photographic and DNA evidence from Melville Island, Northwest Territories. *Arctic* 60: 271–276.

Durner, G.M., and Amstrup, S.C. 1995. Movements of a polar bear from northern Alaska to northern Greenland. *Arctic* 48(4): 338–341.

Durner, G.M. and Amstrup, S.C. 1996. Mass and body-dimension relationships of polar bears in northern Alaska. *Wildlife Society Bulletin* 24(3): 480–484.

Dyck, M., Campbell, M., Lee, D., Boulanger, J. and Hedman, D. 2017. 2016 Aerial survey of the Western Hudson Bay polar bear subpopulation. Final report, Nunavut Department of Environment, Wildlife Research Section, Iglolik, NU.

Escajeda, E., Laidre, K.L., Born, E.W., Wiig, O, Atkinson, S., Dyck, M., Ferguson, S.H. and Lunn, N.J. 2018. Identifying shifts in maternity den phenology and habitat characteristics of polar bears (*Ursus maritimus*) in Baffin Bay and Kane Basin. *Polar Biology* 41: 87–100.

Fauchald, P., Arneberg, P., Berge, J., Gerland, S., Kovacs, K.M., Reigstad, M. and Sundet, J.H. 2014. An assessment of MOSJ - the state of the marine environment around Svalbard and Jan Mayen. Norwegian Polar Institute Report Series no. 145. Available at http://www.mosj.no/en /documents/ [accessed 15 February 2017]

Federal Register announcement (USFWS; 4 October 2017) Endangered and Threatened Wildlife and Plants; 12-Month Findings on Petitions to List 25 Species as Endangered or Threatened Species.

Fedoseev, G.A. 1975. Ecotypes of the ringed seal (Pusa hispida Schreber, 1777) and their reproductive capabilities. In *Biology of the Seal*, K. Ronald and A.W. Mansfield (eds.), pp. 156–160. Rapports et Proces-verbaux des Reunions, Conseil International Pour L'Exploration de la Mer 169.

Ferguson, S.H., Stirling, I. and McLoughlin, P. 2005. Climate change and ringed seal (*Phoca hispida*) recruitment in Western Hudson Bay. *Marine Mammal Science* 21: 121–135.

Finley, K.J, Miller, G.W., Davis, R.A., and Koski, W.R. 1983. A distinctive large breeding population of ringed seal (*Phoca hispida*) inhabiting the Baffin Bay pack ice. *Arctic* 36: 162–173.

Fleck, S. and Hcrrero, S.H. 1988. *Polar bear-human conflicts*. Contract report prepared for Parks Canada and Department of Renewable Resources, Northwest Territories.

Folger, T. 2018. Last ice. *National Geographic* 233: 90–109.

Frasier, T.R., Gillett, R.M., Hamilton, P.K., Brown, M.W., Kraus, S.D. and White, B.N. 2013. Postcopulatory selection for dissimilar gametes maintains heterozygosity in the endangered North Atlantic right whale. *Ecology and Evolution* 3(10): 3483–3494.

Furnell, D.J. and Schweinsburg, R.E. 1984. Population-dynamics of central Canadian Arctic island polar bears. *Journal of Wildlife Management* 48: 722–728.

Gabrielson, K.M., Krokstad, J.S., Villanger, G.D., Blair, D.A.D., Obregon, M.-J., Sonne, C., Dietz, R., Letcher, R.J. and Jenssen, B.M. 2015. Thyroid hormones and deiodinase activity in plasma and tissues in relation to high levels of organohalogen contaminants in East Greenland polar bears (*Ursus maritimus*). *Environmental Research* 136: 413–423.

Gjertz. I ., Aarvik, S. and Hindrum, R. 1993. Polar bears killed in Svalbard 1987–1992. *Polar Research* 12: 107–109.

Gjertz. I . and Persen. E. 1987. Confrontations between humans and polar bears in Svalbard. *Polar Research* 5: 253–256.

Gormezano, L.J. and Rockwell, R.F. 2013a. What to eat now? Shifts in polar bear diet during the ice-free season in western Hudson Bay. *Ecology and Evolution* 3: 3509–3523. DOI: 10.1002/ece3.740.

Gormezano, L.J. and Rockwell, R.F. 2013b. Dietary composition and spatial patterns of polar bear foraging on land in western Hudson Bay. *BMC Ecology* 13: 51. DOI: 10.1186/1472–6785–13–51.

Gosselin, M., Levasseur, M., Wheeler, P.A., Horner, R.A., and Booth, B.C. 1997. New measurements of phytoplankton and ice algae production in the Arctic Ocean. *Deep-Sea Research II* 44: 1623–1644.

Griffen, B.D. 2018. Modeling the metabolic costs of swimming in polar bears (*Ursus maritimus*). Polar Biology. DOI 10.1007/s00300–017–2209-x

Hammill, M.O. and Smith T.G. 1991. The role of predation in the ecology of the ringed seal in BarrowStrait, Northwest Territories, Canada. Marine Mammal Science 7: 123–135.

Harris, R.B. 2012. Preface to 'A circumpolar monitoring framework for polar bears.' *Ursus* 23 (sp2): fmiii–fmiii.

Harwood, L.A., Smith, T.G. and Melling, H. 2000. Variation in reproduction and body condition of the ringed seal (*Phoca hispida*) in western Prince Albert Sound, NT, Canada, as assessed through a harvest-based sampling program. *Arctic* 53: 422–431.

Harwood, L.A., Smith, T.G., Melling, H., Alikamik, J. and Kingsley, M.C.S. 2012. Ringed seals and sea ice in Canada's western Arctic: harvest-based monitoring 1992–2011. *Arctic* 65: 377–390.

Harwood, L.A., Smith, T.G., George, J.C., Sandstrom, S.J., Walkusz, W. and Divoky, G.J. 2015. Change in the Beaufort Sea ecosystem: Diverging trends in body condition and/or production in five marine vertebrate species. *Progress in Oceanography* 136: 263-273.

Herreman, J. and Peacock, E. 2013. Polar bear use of a persistent food subsidy: Insights from non-invasive genetic sampling in Alaska. *Ursus* 24(2): 148–163.

Hoelzel, A.R., Halley, J., O'Brien, S.J., Campagna, C., Arnbom, T., Le Boeuf, B., Ralls, K., and Dover, G.A. 1993. Elephant seal genetic variation and the use of simulation models to investigate historical population bottlenecks. *Journal of Heredity* 84: 443–449.

Honderich, J.E. 1991. Wildlife as a hazardous resource: an analysis of the historical interaction of humans and polar bears in the Canadian arctic. MA thesis, University of Waterloo, Ontario.

Hunter, C.M., Caswell, H., Runge, M.C., Regehr, E.V., Amstrup, S.C. and Stirling, I. 2007. *Polar bears in the Southern Beaufort Sea II: Demography and population growth in relation to sea ice conditions.* US Geological Survey. Reston, Virginia.

Hunter, C.M., Caswell, H., Runge, M.C., Regehr, E.V., Amstrup, S.C. and Stirling, I. 2010. Climate change threatens polar bear populations: a stochastic demographic analysis. *Ecology* 91: 2883–2897.

Iles, D.T., Peterson, S.L., Gormezano, L.J., Koons, D.N. and Rockwell, R.F. 2013. Terrestrial predation by polar bears: not just a wild goose chase. *Polar Biology* 36: 1373–1379.

International Human-Bear Conflicts Workshop. 2009. Polar bear focus day summary (compiled by Colleen Matt). November 17, Canmore, Alberta.

Iverson, S.A., Gilchrist, H.G., Smith, P.A., Gaston, A.J. and Forbes, M.R. 2014. Longer ice-free seasons increase the risk of nest depredation by polar bears for colonial breeding birds in the Canadian Arctic. *Proceedings of the Royal Society B* 281(1779): 20133128.

Johnson, A.C., Pongracz, J.D. and Derocher, A.E. 2017. Long-distance movement of a female polar bear from Canada to Russia. *Arctic* 70(2): 121–128.

Kaeuffer, R., Coltman, D.W., Chapuis, J.-L., Pontier, D., and Réale, D. 2007. Unexpected heterozygosity in an island mouflon population founded by a single pair of individuals. *Proceedings of the Royal Society B* 274: 527–533.

Kelly, B.P., Bengtson, J.L., Boveng, P.L., Cameron, M.F., Dahle, S.P., Jansen, J.K., Logerwell, E.A., Overland, J.E., Sabine, C.L., Waring, G.T. and Wilder, J.M. 2010. Status review of the ringed seal (*Phoca hispida*). NOAA Technical Memorandum NMFS-AFSC-212.

Kelly, B., Whiteley, A., and Tallmon, D. 2010. Comment: The Arctic melting pot. *Nature* 468: 891.

Knudsen, B. 1978. Time budgets of polar bears (*Ursus maritimus*) on North Twin Island, James Bay, during summer. *Canadian Journal of Zoology* 56(7): 1627–1628.

Kolenosky, G.B. and Prevett, J.P. 1983. Productivity and maternity denning of polar bears in Ontario. *Bears: Their Biology and Management* 5: 238–245.

Koop, B.F., Burbidge, M., Byun, A., Rink, U. and Crockford, S.J. 2000. Ancient DNA evidence of a separate origin for North American indigenous dogs. pp. 271–286. *In:* S.J. Crockford (ed.) *Dogs Through Time: An Archaeological Perspective.* Oxford, U.K., Archaeopress S889.

Kovacs, K.M. 2015. *Pagophilus groenlandicus*. The IUCN Red List of Threatened Species 2015: e.T41671A45231087. http://dx.doi.org/10.2305/IUCN.UK.2015--4.RLTS.T41671A45231087.en.

Kovacs, K.M. 2016a. *Cystophora cristata*. The IUCN Red List of Threatened Species 2016: e.T6204A45225150. http://dx.doi.org/10.2305/IUCN.UK.2016--1.RLTS.T6204A45225150.en.

Kovacs, K.M. 2016b. *Erignathus barbatus*. The IUCN Red List of Threatened Species 2016: e.T8010A45225428. Available online http://www.iucnredlist.org/details/full/8010/0.

Kumar, V., Lammers, F., Bidon, T., Pfenninger, M., Kolter, L., Nilsson, M.A., and Janke, A. 2017. The evolutionary history of bears is characterized by gene flow across species. *Scientific Reports* 7: 46487.

Laidre, K.L., Born, E.W., Atkinson, S.N., Wiig, Ø., Andersen, L.W., Lunn, N.J., Dyck, M., Regehr, E.V., McGovern, R. and Heagerty, P. 2018. Range contraction and increasing isolation of a polar bear subpopulation in an era of sea-ice loss. *Ecology and Evolution* DOI: 10.1002/ece3.3809.

Laidre, K.L., Regehr, E.V., Akcakaya, H.R., Amstrup, S.C., Atwood, T., Lunn, N., Obbard, M., Stern, H.L., Thiemann, G. and Wiig, Ø. 2016. AGU 2016. Fall Meeting, San Francisco, presentation of Regehr et al. 2016 paper results. 12 December 2016. https://agu.confex.com/agu/fm16/meetingapp.cgi/Paper/119105. 'GC13I-01: Conservation status of polar bears (Ursus maritimus) in relation to projected sea-ice declines (Invited)'

Lang, A., Yang, S. and Kaas, E. 2017. Sea ice thickness and recent Arctic warming *Geophysical Research Letters.* DOI: 10.1002/2016GL071274

Larsen, T. 1972. Norwegian polar bear hunt, management and research. *Bears: Their Biology and Management* 2: 159–164. http://www.jstor.org/stable/3872579.

Larsen, T. 1985. Polar bear denning and cub production in Svalbard, Norway. *Journal of Wildlife Management* 49: 320–326.

Lee, S.H. and Whitledge, T.E. 2005. Primary and new production in the deep Canada Basin during summer 2002. *Polar Biology* 28: 190–197.

Lehman N, 1998. Genes are not enough. *Current Biology* 8: 1078–1081 [R722-R724].

Lindqvist, C., Schuster, S.C., Sun, Y., Talbot, S.L., Qi, J., Ratan, A., Tomsho, L., Kasson, L., Zeyl, E., Aars, J., Miller, W., Ingólfsson, Ó., Bachmann, L. and Wiig, Ø. 2010. Complete mitochondrial genome of a Pleistocene jawbone unveils the origin of polar bear. *Proceedings of the National Academy of Sciences USA* 107: 5053–5057.

Lowry, L. 2015. *Odobenus rosmarus ssp. divergens*. The IUCN Red List of Threatened Species 2015, e.T61963499A45228901.
Online at http://www.iucnredlist.org/details/full/61963499/0.

Lowry, L. 2016. *Pusa hispida*. The IUCN Red List of Threatened Species 2016: e.T41672A45231341. Available online http://www.iucnredlist.org/details/full/61382318/0.

Lunn, N.J., Schliebe, S., and Born, E.W. (eds.). 2002. Polar Bears: Proceedings of the 13th meeting of the Polar Bear Specialist Group IUCN/SSC, 23–28 June 2001, Nuuk, Greenland. Gland, Switzerland and Cambridge UK, IUCN.

Lunn, N.J. and Stenhouse, G.B. 1985. An observation of possible cannibalism by polar bears (*Ursus maritimus*). *Canadian Journal of Zoology* 63: 1516–1517.

Matishov, G.G., Chelintsev, N.G., Goryaev, Yu. I., Makarevich, P.R. and Ishkulov, D.G. 2014. Assessment of the amount of polar bears (*Ursus maritimus*) on the basis of perennial vessel counts. *Doklady Earth Sciences* 458 (2): 1312–1316.

MacCracken, J.G., Beatty, W.S., Garlich-Miller, J.L., Kissling, M.L and Snyder, J.A. 2017. *Final Species Status Assessment for the Pacific Walrus* (*Odobenus rosmarus divergens*), May 2017 (Version 1.0). US Fish & Wildlife Service, Anchorage, AK.

McKinney, M.A., Iverson, A.J., Fisk, A.T., Sonne, C., Rigét, F.F., Letcher, R.J., Arts, M.T., Born, E.W., Rosing-Asvid, A. and Dietz, R. 2013. Global change effects on the long-term feeding ecology and contaminant exposures of East Greenland polar bears. *Global Change Biology* 19: 2360–2372

McLoughlin, P.D., Taylor, M.K., Cluff, H.D., Gau, R.J., Mulders, R., Case, R.L. and Messier, F. 2003. Population viability of barren-ground grizzly bears in Nunavut and the Northwest Territories. *Arctic* 56: 185–190.

Miller, S., Schliebe, S. and Proffitt, K. 2006. Demographics and Behavior of Polar Bears Feeding on Bowhead Whale Carcasses at Barter and Cross Islands, Alaska, 2002–2004. Alaska Outer Continental Shelf (OCS) Study MMS 2006–14, US Dept. of the Interior, Minerals Management Service, Anchorage.

Miller, W., Schuster, S.C., Welch, A.J., Ratan, A., Bedoya-Reina, O.C., Zhao, F., Kim, H.L., Burhans, R.C., Drautz, D.I., Wittekindt, N.E., Tomsho, L.P., Ibarra-Laclette, E., Herrera-Estrella, L., Peacock, E., Farley, S., Sage, G.K., Rode, K., Obbard, M., Montiel, R., Bachmann, L., Ingolfsson, O., Aars, J., Mailund, T., Wiig, Ø., Talbot, S.L., and Lindqvist, C. 2012. Polar and brown bear genomes reveal ancient admixture and demographic footprints of past climate change. *Proceedings of the National Academy of Sciences* 109(36): E2382-E2390.

Molnar, P.K., Derocher, A.E., Theimann, G., and Lewis, M.A. 2010. Predicting survival, reproduction and abundance of polar bears under climate change. *Biological Conservation* 143: 1612–1622.

Monnett, C. and Gleason, J.S. 2006. Observations of mortality associated with extended open-water swimming by polar bears in the Alaskan Beaufort Sea. *Polar Biology* 29: 681–687.

Moore, C.R., West, A., LeCompte, M.A., Brooks, M.J., Daniel Jr., I.R., Goodyear, A.C., Ferguson, T.A., Ivester, A.H., Feathers, J.K., Kennett, J.P., Tankersley, K.B., Adedeji, A.V. and Bunch, T.E. 2017. Widespread platinum anomaly documented at the Younger Dryas onset in North American sedimentary sequences. *Nature Scientific Reports* 7: 44031.

Nielsen, S.E., Cattet, M.R.L, Boulanger, J., Cranston, J., McDermid, G.J., Shafer, A.B.A. and Stenhouse, G.B. 2013. Environmental, biological and anthropogenic effects on grizzly bear body size: temporal and spatial considerations. *BMC Ecology* 13: 31.

Nowak, R.M. 2003. *Walker's Marine Mammals of the World*. Johns Hopkins University Press, Baltimore.

Nyugen, L., Pilfold, N.W., Derocher, A.E., Stirling, I., Bohart, A.M., and Richardson, E. 2017. Ringed seal (*Pusa hispida*) tooth annuli as an index of reproduction in the Beaufort Sea. *Ecological Indicators* 77: 286–292.

Obbard, M.E. 2008. Southern Hudson Bay polar bear project 2003–2005: Final report. Unpublished report, Wildlife Research and Development Section, Ontario Ministry of Natural Resources, Peterborough, ON.

Obbard, M.E., Cattet, M.R.L., Moody, T. Walton, L.R., Potter, D, Inglis, J. and Chenier, C. 2006. Temporal trends in the body condition of Southern Hudson Bay polar bears. *Climate Change Research Information Note, No. 3*. Ontario Ministry of Natural Resources, Applied Research and Development Branch, Sault Ste. Marie, Ontario, Canada.

Obbard, M.E., Cattet, M.R.L., Howe, E.J., Middel, K.R., Newton, E.J., Kolenosky, G.B., Abraham, K.F. and Greenwood, C.J. 2016. Trends in body condition in polar bears (*Ursus maritimus*) from the Southern Hudson Bay subpopulation in relation to changes in sea ice. *Arctic Science* 2: 15–32.

Obbard, M.E., Stapleton, S., Middel, K.R., Thibault, I., Brodeur, V. and Jutras, C. 2015. Estimating the abundance of the Southern Hudson Bay polar bear subpopulation with aerial surveys. *Polar Biology* 38: 1713–1725.

Obbard, M.E., Theimann, G.W., Peacock, E. and DeBruyn, T.D. (eds.) 2010. Polar Bears: Proceedings of the 15th meeting of the Polar Bear Specialist Group IUCN/SSC, 29 June-3 July, 2009, Copenhagen, Denmark. Gland, Switzerland and Cambridge UK, IUCN.

Olesiuk, P.F., Bigg, M.A., Ellis, G.M., Crockford, S.J. and Wigen, R.J. 1990. An assessment of the feeding habits of harbour seals (Phoca vitulina) in the Strait of Georgia, British Columbia, based on scat analysis. *Canadian Technical Reports on Fisheries and Aquatic Sciences* 1730.

Olson, J.W., Rode, K.D., Eggett, D., Smith, T.S., Wilson, R.R., Durner, G.M., Fischbach, A., Atwood, T.C. and Douglas, D.C. 2017. Collar temperature sensor data reveal long-term patterns in Southern Beaufort Sea polar bear den distribution on pack ice and land. *Marine Ecology Progress Series* 564: 211–224.

Overland, J.E. and Wang, M. 2013. When will the summer Arctic be nearly sea ice free? *Geophysical Research Letters* 40: 2097–2101.

Ovodov, N.D., Crockford, S.J., Kuzmin, Y.V., Higham, T.F.G., Hodgins, G.W.L., and van der Plicht, J. 2011. A 33,000 year old incipient dog from Altai Mountains of Siberia: evidence for the earliest domestication disrupted by the Last Glacial Maximum. *PLoS One* 6(7): e22821.

Ovsyanikov, N. 2010. Polar bear research on Wrangel Island and in the central Arctic Basin. In, Proceedings of the 15th meeting of the Polar Bear Specialist Group IUCN/SSC, 29 June-3 July, 2009, edited by Obbard, M.E., Theimann, G.W., Peacock, E. and DeBryn, T.D., pp. 171–178. Gland, Switzerland and Cambridge UK, IUCN.

Ovsyanikov, N.G. and Menyushina, I.E. 2015. Demographic processes in Chukchi-Alaskan polar bear population as observed in Wrangel Island region. pp. 37–55, In: *Marine Mammals of the Holarctic,* Collection of Scientific Papers. Vol. 2. Moscow.

Pomeroy, L.R. 1997. Primary production in the Arctic Ocean estimated from dissolved oxygen. *Journal of Marine Systems* 10: 1–8.

Packard, A.S. 1886. The former southern limits of the white or polar bear. *American Naturalist* 20(7): 655–659.

Pagano, A.M., Durner, G.M., Amstrup, S.C., Simac, K.S. and York, G.S. 2012. Long-distance swimming by polar bears (*Ursus maritimus*) of the Southern Beaufort Sea during years of extensive open water. *Canadian Journal of Zoology* 90: 663–676.

Pagano, A.M., Durner, G.M., Rode, K.D., Atwood, T.C., Atkinson, S.N., Peacock, E., Costa, D.P., Owen, M.A. and Williams, T.M. 2018. High-energy, high-fat lifestyle challenges an Arctic apex predator, the polar bear. *Science* 359 (6375): 568.

Peacock, E., Taylor, M.K., Laake, J., and Stirling, I. 2013. Population ecology of polar bears in Davis Strait, Canada and Greenland. *Journal of Wildlife Management* 77: 463–476.

Pearce, F. 2018. The survivors: Is climate change really killing polar bears? *New Scientist 3164* (10 February).

PBSG 2017. Summary of polar bear population status per 2017. Update 30 March 2017. http://pbsg.npolar.no/en/status/status-table.html [assessed 17 January 2018]

Perovich, D., Meier, W., Tschudi, M., Farrell, S., Hendricks, S., Gerland, S., Haas, C., Krumpen, T., Polashenski, C., Ricker, R. and Webster, M. 2017. Sea ice. *Arctic Report Card 2017.*, NOAA.

Pilfold, N.W., Derocher, A.E., Stirling, I. and Richardson, E. 2014. Polar bear predatory behaviour reveals seascape distribution of ringed seal lairs. *Population Ecology* 56: 129–138.

Pilfold, N.W., Derocher, A.E., Stirling, I. and Richardson, E. 2015. Multi-temporal factors influence predation for polar bears in a changing climate. *Oikos* 124 (8): 1098–1107.

Pilfold, N.W., Derocher, A.E., Stirling, I., Richardson, E. and Andriashek, D. 2012. Age and sex composition of seals killed by polar bears in the Eastern Beaufort Sea. *PLoS ONE* 7(7): e41429.

Pilfold, N.W., McCall, A., Derocher, A.E., Lunn, N.J., and Richardson, E. 2017. Migratory response of polar bears to sea ice loss: to swim or not to swim. *Ecography* 40: 189–199.

Pongracz, J.D., and Derocher, A.E. 2017. Summer refugia habitat of polar bears (*Ursus maritimus*) in the southern Beaufort Sea. *Polar Biology* 40(4): 753–763.

Pongracz, J.D., Paetkau, D., Branigan, M. and Richardson, E. 2017. Recent hybridization between a polar bear and grizzly bears in the Canadian Arctic. *Arctic* 70: 151–160.

Post, E., Bhatt, U.S., Bitz, C.M., Brodie, J.F., Fulton, T.L., Hebblewhite, M., Kerby, J., Kutz, S.J., Stirling, I. and Walker, D.A. 2013. Ecological consequences of sea-ice decline. *Science* 341: 519–524.

Rajan, A. and Tol, R.S. 2018. Lipstick on a bear: A comment on internet blogs, polar bears, and climate-change denial by proxy. *EarthArXiv*. eartharxiv.org/ct84n. *BioScience*, submitted.

Ramsay, M.A. and Stirling, I. 1988. Reproductive biology and ecology of female polar bears (*Ursus maritimus*). *Journal of Zoology London* 214: 601–624.

Ramseier, R.O., Vant, M.R., Arsenault, L.D., Gray, L., Gray, R.B., and Chudobiak, W.J. 1975. Distribution of the ice thickness in the Beaufort Sea. *Beaufort Sea Technical Report #30*. Canada Dept. of Environment, Victoria, B.C. Available online.

Regehr, E.V., Amstrup, S.C., and Stirling, I. 2006. *Polar bear population status in the Southern Beaufort Sea.* US Geological Survey Open-File Report 2006–1337.

Regehr, E.V., Lunn, N.J., Amstrup, S.C. and Stirling, I. 2007. Effects of earlier sea ice breakup on survival and population size of polar bears in Western Hudson Bay. *Journal of Wildlife Management* 71: 2673–2683.

Regehr, E.V., Laidre, K.L, Akçakaya, H.R., Amstrup, S.C., Atwood, T.C., Lunn, N.J., Obbard, M., Stern, H., Thiemann, G.W., and Wiig, Ø. 2016. Conservation status of polar bears (*Ursus maritimus*) in relation to projected sea-ice declines. *Biology Letters* 12: 20160556.

Rode, K.D., Amstrup, S.C. and Regehr, E.V. 2010. Reduced body size and cub recruitment in polar bears associated with sea ice decline. *Ecological Applications* 20(3): 768–782.

Rode, K.D., Douglas, D., Durner, G., Derocher, A.E., Thiemann, G.W., and Budge, S. 2013. *Variation in the response of an Arctic top predator experiencing habitat loss: feeding and reproductive ecology of two polar bear populations*. Oral presentation by Karyn Rode, 28th Lowell Wakefield Fisheries Symposium, March 26–29. Anchorage, AK.

Rode, K.D., Olson, J., Eggett, D., Douglas, D.C., Durner, G.M., Regehr, E.V., Wilson, R.R., Smith, T. and St. Martin, M. 2018a. Den phenology and reproductive success of polar bears in a changing climate. *Journal of Mammalogy* DOI: 10.1093/jmammal/gyx181

Rode, K.D., Wilson, R.R., Douglas, D.C., Muhlenbruch, V., Atwood, T.C., Regehr, E.V., Richardson, E.S., Pilfold, N.W., Derocher, A.E., Durner, G.M., Stirling, I., Amstrup, S.C., St. Martin, M., Pagano, A.M. and Simac, K. 2018b. Spring fasting behavior in a marine apex predator provides an index of ecosystem productivity. *Global Change Biology*

Rode, K.D., Peacock, E., Taylor, M., Stirling, I., Born, E.W., Laidre, K.L., and Wiig, Ø. 2012. A tale of two polar bear populations: ice habitat, harvest, and body condition. *Population Ecology* 54: 3–18.

Rode, K.D. and Regehr, E.V. 2010. Polar bear research in the Chukchi and Bering Seas: A synopsis of 2010 field work. Unpublished report to the US Fish and Wildlife Service, Department of the Interior, Anchorage.

Rode, K.D., Regehr, E.V., Douglas, D., Durner, G., Derocher, A.E., Thiemann, G.W., and Budge, S. 2014. Variation in the response of an Arctic top predator experiencing habitat loss: feeding and reproductive ecology of two polar bear populations. *Global Change Biology* 20(1): 76–88.

Rode, K.D., Robbins, C.T., Nelson, L. and Amstrup, S.C. 2015. Can polar bears use terrestrial foods to offset lost ice-based hunting opportunities? *Frontiers in Ecology and the Environment* 13(3): 138–145, doi: 10.1890/140202

Rogers, M.C., Peacock, E., Simac, K., O'Dell, M.B. and Welker, J.M. 2015. Diet of female polar bears in the southern Beaufort Sea of Alaska: evidence for an emerging alternative foraging strategy in response to environmental change. *Polar Biology* 38(7): 1035–1047.

Rolland, N. and Crockford, S.J. 2005. Late Pleistocene dwarf *Stegodon* from Flores, Indonesia? *Antiquity* 79 (June #304 Project Gallery). http://antiquity/ac.uk/projgall/rolland/.

Sergeant, D.E. 1976. History and present status of populations of harp and hooded seals. *Biological Conservation* 10: 95–118.

Sergeant, D.E. 1991. *Harp Seals, Man and Ice.* Canadian Special Publication of Fisheries and Aquatic Sciences 114. Fisheries and Oceans Canada, Ottawa.

Sha, L., Jiang, H., Seidenkrantz, M-S., Muscheler, R., Zhang, X., Knudsen, M.F., Olsen, J., Knudsen, K.L. and Zhang, W. 2016. Solar forcing as an important trigger for West Greenland sea-ice variability over the last millennium. *Quaternary Science Reviews* 131: 148–156.

Smith, T.G. 1987. The ringed seal, *Phoca hispida*, of the Canadian Western Arctic. *Canadian Bulletin of Fisheries and Aquatic Sciences 216*. Department of Fisheries and Oceans, Ottawa.

Smith, T.G. and Hammill, M.O. 1981. Ecology of the ringed seal, *Phoca hispida*, in its fast ice breeding habitat. *Canadian Journal of Zoology* 59: 966–981.

Smith, T.G., Hammill, M.O. and Taugbøl, G. 1991. A review of the development, behavioural and physiological adaptations of the ringed seal, *Phoca hispida*, to life in the Arctic winter. *Arctic* 44: 124–131.

Smith, T.G. and Stirling, I. 1978. Variation in the density of ringed seal (*Phoca hispida*) birth lairs in the Amundsen Gulf, Northwest Territories. *Canadian Journal of Zoology* 56: 1066–1071.

Smith, P., Stirling, I., Jonkel, C., and Juniper, I. 1975. Notes on the present status of the polar bear (*Ursus maritimus*) in Ungava Bay and northern Labrador. *Canadian Wildlife Service, Progress Notes 53.*

Sonne C. 2010. Health effects from long-range transported contaminants in Arctic top predators: an integrated review based on studies of polar bears and relevant model species. *Environment International* 36: 461–91.

Sonne, C., Dyck, M., Rigét, F.F., Jensen, J.-E.B., Hyldstrup, L., Letcher, R.J., Gustavson, K., Gilbert, M.T.P., Dietz, R. 2015. Penile density and globally used chemicals in Canadian and Greenland polar bears. *Environmental Research* 137: 287–291.

Stapleton S., Atkinson, S., Hedman, D., and Garshelis, D. 2014. Revisiting Western Hudson Bay: using aerial surveys to update polar bear abundance in a sentinel population. Biological Conservation 170: 38–47.

Stapleton S., Peacock, E., and Garshelis, D. 2016. Aerial surveys suggest long-term stability in the seasonally ice-free Foxe Basin (Nunavut) polar bear population. *Marine Mammal Science* 32: 181–201.

Stein, R., Fahl, K., Schade, I., Manerung, A., Wassmuth, S., Niessen, F. and Nam, S-I. 2017. Holocene variability in sea ice cover, primary production, and Pacific-Water inflow and climate change in the Chukchi and East Siberian Seas (Arctic Ocean). *Journal of Quaternary Science* 32: 362–379.

Stenhouse, G.B., Lee, L.J. and Poole K.G. 1988. Some characteristics of polar bears killed during conflicts with humans in the Northwest Territories, 1976–86. *Arctic* 41: 275–278.

Stenson, G.B. 2014. *The status of harp and hooded seals in the North Atlantic.* Report presented at the Scientific Council Meeting, June 2014. Northwest Atlantic Fisheries Organization SCR Doc. 14/026, Serial No. N6321.

Stenson, G.B., Buren, A.D. and Koen-Alonso, M. 2015. The impact of changing climate and abundance on reproduction in an ice-dependent species, the Northwest Atlantic harp seal, *ICES Journal of Marine Science* 73(2): 250–262.

Stern, H.L. and Laidre, K.L. 2016. Sea-ice indicators of polar bear habitat. *The Cryosphere* doi: 10.5194/tc-10-2027-2016.

Stirling, I. 1974. Midsummer observations on the behavior of wild polar bears (*Ursus maritimus*). *Canadian Journal of Zoology* 52: 1191–1198.

Stirling, I. 2002. Polar bears and seals in the eastern Beaufort Sea and Amundsen Gulf: a synthesis of population trends and ecological relationships over three decades. *Arctic* 55 (Suppl. 1): 59–76.

Stirling, I. and Archibald, R. 1977. Aspects of predation of seals by polar bears. *Journal of the Fisheries Research Board of Canada* 34: 1126–1129.

Stirling, I. and Derocher, A.E. 2007. Melting Under Pressure. *The Wildlife Professional,* Fall: 24–27, 43.

Stirling, I. and Derocher, A.E. 2012. Effects of climate warming on polar bears: a review of the evidence. *Global Change Biology* 18: 2694–2706.

Stirling, I. and Kiliaan, H.P. 1980. Population ecology studies of the polar bear in northern Labrador. *Canadian Wildlife Service Occasional Paper No. 42.*

Stirling, I. and Lunn, N.J. 1997. Environmental fluctuations in arctic marine ecosystems as reflected by variability in reproduction of polar bears and ringed seals. In *Ecology of Arctic Environments,* Woodin, S.J. and Marquiss, M. (eds), pp. 167–181. Blackwell Science, UK.

Stirling, I. and Øritsland, N.A. 1995. Relationships between estimates of ringed seal (*Phoca hispida*) and polar bear (*Ursus maritimus*) populations in the Canadian Arctic. *Canadian Journal of Fisheries and Aquatic Sciences* 52: 2594–2612.

Stirling, I. and Ross, J.E. 2011. Observations of cannibalism by polar bears (*Ursus maritimus*) on summer and autumn sea ice at Svalbard, Norway. *Arctic* 64: 478–482.

Stirling, I., Andriashek, D., Latour, P.B. and Calvert, W. 1975a. Distribution and abundance of polar bears in the Eastern Beaufort Sea. Beaufort Sea Tech. Report #2, Dept. Environment, Victoria, B.C.

Stirling, I., Archibald, R. and DeMaster, D. 1975b. Distribution and abundance of seals in the Eastern Beaufort Sea. Beaufort Sea Tech. Report #1, Dept. Environment, Victoria, B.C.

Stirling, I., Andriashek, D., Spencer, C. and Derocher, A.E. 1988. Assessment of the polar bear population in the Eastern Beaufort Sea. Final Report to the Northern Oil and Gas Assessment Program. Canadian Wildlife Service, Edmonton, Alberta, Canada.

Stirling, I., Calvert, W., and Andriashek, D. 1984. Polar bear ecology and environmental considerations in the Canadian High Arctic. pp. 201–222. In Olson, R., Geddes, F. and Hastings, R. (eds.). *Northern Ecology and Resource Management.* University of Alberta Press, Edmonton.

Stirling, I, Cleator, H. and Smith, T.G. 1981. Marine mammals. In: Polynyas in the Canadian Arctic, Stirling, I. and Cleator, H. (eds), pp. 45–58. *Canadian Wildlife Service Occasional Paper* No. 45. Ottawa.

Stirling, I, Kingsley, M. and Calvert, W. 1982. The distribution and abundance of seals in the Eastern Beaufort Sea, 1974–79. *Canadian Wildlife Service Occasional Paper* 47. Edmonton.

Stirling, I., Lunn, N.J. and Iacozza, J. 1999. Long-term trends in the population ecology of polar bears in Western Hudson Bay in relation to climate change. *Arctic* 52: 294–306.

Stirling, I., Lunn, N.J., Iacozza, J., Elliott, C., and Obbard, M. 2004. Polar bear distribution and abundance on the southwestern Hudson Bay coast during open water season, in relation to population trends and annual ice patterns. *Arctic* 57: 15–26.

Stirling, I., Richardson, E., Thiemann, G.W. and Derocher, A.E. 2008. Unusual predation attempts of polar bears on ringed seals in the southern Beaufort Sea: possible significance of changing spring ice conditions. *Arctic* 61: 14–22.

Stirling, I., Schweinsburg, R.E., Kolenasky, G.B., Juniper, I., Robertson, R.J., and Luttich, S. 1980. Proceedings of the 7th meeting of the Polar Bear Specialist Group IUCN/SSC, 30 January-1 February, 1979, Copenhagen, Denmark. Gland, Switzerland and Cambridge UK, IUCN., pp. 45–53.

Stirling, I., Schweinsburg, R.E., Kolenasky, G.B., Juniper, I., Robertson, R.J., Luttich, S. and Calvert, W. 1985. Research on polar bears in Canada 1978–80. *In* Polar Bears: Proceedings of the 8th meeting of the Polar Bear Specialist Group IUCN/SSC, Oslo, Norway, 15–19 January 1981. IUCN, Gland, Switzerland. pp. 71–98.

Stroeve, J., Holland, M.M., Meier, W., Scambos, T. and Serreze, M. 2007. Arctic sea ice decline: Faster than forecast. *Geophysical Research Letters* 34:L09501.

Stroeve, J., Markus, T, Boisvert, L., Miller, J. and Barrett, A. 2014. Changes in Arctic melt season and implications for sea ice loss. *Geophysical Research Letters* 41: 1216–1224.

Swaisgood, R.R. and Sheppard, J.K. 2010. The culture of conservation biologists: show me the hope! BioScience 60(8): 626–630.

SWG [Scientific Working Group to the Canada-Greenland Joint Commission on Polar Bear]. 2016. Re-Assessment of the Baffin Bay and Kane Basin Polar Bear Subpopulations: Final Report to the Canada-Greenland Joint Commission on Polar Bear.

SWG Executive Summary [Scientific Working Group to the Canada-Greenland Joint Commission on Polar Bear Executive Summary]. 2016. Re-Assessment of the Baffin Bay and Kane Basin

Polar Bear Subpopulations: Executive Summary for the Final Report to the Canada-Greenland Joint Commission on Polar Bear.

Tartu, S., Bourgeon, S., Aars, J., Andersen, M., Ehrich, D., Thiemann, G.W., Welker, J.M. and Routti, H. 2016. Geographical area and life history traits influence diet in an Arctic marine predator. *PLoS ONE* 11(5): e0155980.

Taylor, K.C., Mayewski, P.A., Alley, R.B., Brook, E.J., Gow, E.J., Grootes, P.M., Meese, D.A., Salzman, E.S., Severinghaus, J.P., Twickler, M.S., White, J.W.C., Whitlow, S. and Zielinski, G.A. 1997. The Holocene-Younger Dryas transition recorded at Summit, Greenland. *Science* 278(5339): 825–827.

Taylor, M.K., Akeeagok, S., Andriashek, D., Barbour, W., Born, E.W., Calvert, W., Cluff, H.D., Ferguson, S., Laake, J., Rosing-Asvid, A., Stirling, I. and Messier, F. 2001. Delineating Canadian and Greenland polar bear (*Ursus maritimus*) populations by cluster analysis of movements. *Canadian Journal of Zoology* 79(4): 690–709.

Taylor, M.K., Laake, J., McLoughlin, P.D., Cluff, H.D., and Messier, F. 2006. Demography parameters and harvest-explicit population viability analysis for polar bears in M'Clintock Channel, Nunavut, Canada. *Journal of Wildlife Management* 70: 1667–1673.

Taylor, M.K., Laake, J., McLoughlin, P.D., Cluff, H.D., and Messier, F. 2008. Mark-recapture and stochastic population models for polar bears of the High Arctic. *Arctic* 61: 143–152.

Taylor, M.K., Laake, J., McLoughlin, P.D., Cluff, H.D., and Messier, F. 2009. Demography and population viability of polar bears in the Gulf of Boothia, Nunavut. *Marine Mammal Science* 25: 778–796.

Taylor, M., Larsen, T. and Schweinsburg, R.E. 1985. Observations of intraspecific aggression and cannibalism in polar bears (Ursus maritimus). *Arctic* 38: 303–309.

Taylor, M., and Lee, J. 1995. Distribution and abundance of Canadian polar bear populations: a management perspective. *Arctic* 48: 147–154.

Todd, F.S., Headland, R.K., and Lasca, N. 1992. Animals at the North Pole. *Polar Record* 28: 321–322.

Tollit, D.J., Schulze, A., Trites, A.W., Olesiuk, P., Crockford, S.J., Gelatt, T., Ream, R., and K. Miller, K. 2009. Development and application of DNA techniques for validating and improving pinniped diet estimates. *Ecological Applications* 19: 889–905.

Truett, J.C. (ed.). 1993. Guidelines for Oil and Gas Operations in Polar Bear Habitats. Minerals Management Service Alaska, US Dept. of the Interior, Report 93–0008.

US Fish & Wildlife Service (USFWS). 2008. Determination of threatened status for the polar bear (*Ursus maritimus*) throughout its range. *Federal Register* 73: 28212–28303.

US Fish & Wildlife Service (USFWS). 2012a. Threatened status for the Arctic, Okhotsk and Baltic subspecies of the ringed seal. *Federal Register* 77: 76706–76738.

US Fish & Wildlife Service (USFWS). 2012b. Threatened status for the Beringia and Okhotsk distinct population segments of the *Erignathus barbatus nauticus* subspecies of the bearded seal. *Federal Register* 77: 76740–76768.

US Fish and Wildlife Service. 2017. Endangered and Threatened Wildlife and Plants; 12-Month Findings on Petitions to List 25 Species as Endangered or Threatened Species.

US Geological Survey (USGS). 2007. Executive Summary, USGS Science Strategy to Support U.S. Fish and Wildlife Service Polar Bear Listing Decision. Administrative Report, US Geological Survey. Reston, Virginia.

Uspenski, S.M. and Kistchinski, A.A. 1972. New data on the winter ecology of the polar bear (Ursus maritimus, Phipps) on Wrangel Island. *Bears: Their Biology and Management* 2: 181–197.

Van de Velde (OMI), F., Stirling, I. and Richardson, E. 2003. Polar bear (*Ursus maritimus*) denning in the area of the Simpson Peninsula, Nunavut. *Arctic* 56: 191–197.

Van Meurs, R. and Splettstoesser, J.F. 2003. Farthest North Polar Bear (Letter to the Editor). *Arctic* 56: 309.

Vibe, C. 1965. The polar bear in Greenland. In, *Proceedings of the First International Scientific Meeting on the Polar Bear.* Fairbanks, Alaska 6–10 September 1965, IUCN Polar Bear Specialist Group. University of Alaska International Conference Proceedings Series, No. 1. pp. 17–25. Washington, DC.

Vibe, C. 1967. *Arctic animals in relation to climatic fluctuations.* Meddelelser om Grønland. 170(5). C.A. Reitzels Forlag, Copenhagen.

Vongraven, D. and Peacock E. 2011. Development of a pan-Arctic monitoring plan for polar bears: background paper. Circumpolar Biodiversity Monitoring Programme, *CAFF Monitoring Series Report No. 1*, CAFF International Secretariat, Akureyri, Iceland. Available at http://www.caff.is/publications/

Vongraven, D., Aars, J., Amstrup, S., Atkinson, S.N., Belikov, S., Born, E.W., DeBruyn, T., Derocher, A.E., Durner, G., Gill, M., Lunn, N., Obbard, M., Omelak, J., Ovsyanikov, N., Peacock, E., Richardson, E., Sahanatien, V., Stirling, I. and Wiig, Ø. 2012. A circumpolar monitoring framework for polar bears. *Ursus 23 (sp2)*: 1–66.

Vongraven D. 2013. Circumpolar monitoring framework for polar bears. Presentation 5.1 at the International Polar Bear Forum, December 3–6, Moscow.

Walsh, J.E., Fetterer, F., Stewart, J.S. and Chapman, W.L. 2017. A database for depicting Arctic sea ice variations back to 1850. *Geographical Review* 107(1): 89–107.

Wang, M. and Overland, J.E. 2012. A sea ice free summer Arctic within 30 years: An update from CMIP5 models. *Geophysical Research Letters* 39: L18501.

Wang, M. and Overland, J.E. 2015. Projected future duration of the sea-ice-free season in the Alaskan Arctic. *Progress in Oceanography* 136: 50–59.

Weber, D.S., Stewart, B.S., Garza, J.C., and Lehman, N. 2000. An empirical genetic assessment of the severity of the northern elephant seal population bottleneck. *Current Biology* 10: 1287–1290.

Wiig, Ø., Derocher, A.E. and Belikov, S.E. 1999. Ringed seal (*Phoca hispida*) breeding in the drifting pack ice of the Barents Sea. *Marine Mammal Science* 15: 595– 598.

Wiig, Ø., Amstrup, S., Atwood, T., Laidre, K., Lunn, N., Obbard, M., et al. 2015. *Ursus maritimus. The IUCN Red List of Threatened Species 2015.*: e.T22823A14871490. Available from http://www.iucnredlist.org/details/22823/0 [accessed Nov. 28, 2015]

Wilder, J.M., Vongraven, D., Atwood, T., Hansen, B., Jessen, A., Kochnev, A., York, G., Vallender, R., Hedman, D. and Gibbons, M. 2017. Polar bear attacks on humans: implications of a changing climate. *Wildlife Society Bulletin* DOI: 10.1002/wsb.783.

York, J., Dowsley, M., Cornwell, A., Kuc, M. and Taylor, M. 2016. Demographic and traditional knowledge perspectives on the current status of Canadian polar bear subpopulations. *Ecology and Evolution* 6(9): 2897–2924.

About the Global Warming Policy Foundation

The Global Warming Policy Foundation is an all-party and non-party think tank and a registered educational charity which, while openminded on the contested science of global warming, is deeply concerned about the costs and other implications of many of the policies currently being advocated.

Our main focus is to analyse global warming policies and their economic and other implications. Our aim is to provide the most robust and reliable economic analysis and advice. Above all we seek to inform the media, politicians and the public, in a newsworthy way, on the subject in general and on the misinformation to which they are all too frequently being subjected at the present time.

The key to the success of the GWPF is the trust and credibility that we have earned in the eyes of a growing number of policy makers, journalists and the interested public. The GWPF is funded overwhelmingly by voluntary donations from a number of private individuals and charitable trusts. In order to make clear its complete independence, it does not accept gifts from either energy companies or anyone with a significant interest in an energy company.

Views expressed in the publications of the Global Warming Policy Foundation are those of the authors, not those of the GWPF, its trustees, its Academic Advisory Council members or its directors.

Made in the USA
Monee, IL
05 November 2026

67383064R00043